COMPUTER
	COMMUNICATIONS
	TECHNOLOGY

October 3, 1983

Kenneth Beaman
114 Pine Wood Lane
Los Gatos, CA 95030

Enclosed is your order for the "Solution Manual" for Queueing Systems Volume I.

Your cancelled check is your receipt.

Thank you for your order.

D.M. Kaye
TTI Publications Dept.

741 10th Street / Santa Monica, CA 90402 / (213) 394-8305

SOLUTIONS MANUAL

FOR

QUEUEING SYSTEMS

VOLUME I: THEORY

SOLUTIONS MANUAL

FOR

QUEUEING SYSTEMS
VOLUME I: THEORY

Leonard Kleinrock
Richard Gail

Computer Science Department
School of Engineering and Applied Science
University of California, Los Angeles

SANTA MONICA, CA 90402

Copyright © 1982 by Technology Transfer Institute
Santa Monica, California 90402

All rights reserved. No part of this
book may be reproduced in any form
or by any means without permission
in writing from the publisher.

Library of Congress Catalog Card Number: 82-80907
ISBN 0-942948-00-9

Printed in the United States of America

10 9 8 7 6 5 4 3 2 1

Preface

In your hands you are holding the long awaited complete solutions manual to my text on queueing systems theory which first appeared in the marketplace six years ago. This manual has taken too long to prepare and its appearance has been delayed more often than I care to admit — so much for queueing theory expertise. In fact, there are those among us who were convinced that this solutions manual was simply a figment of my wishful thinking.

In the time since its publication, this text has been thoroughly tested in the harshest possible environment, namely the graduate program in Computer Science at UCLA. A number of other universities have also adopted this text for their queueing theory offerings. In addition to its adoption as a university text, this book has served as a reference and self-teaching guide for the professional world of engineers, operations researchers, scientists, and, would you believe, even mathematicians. Hundreds of critical readers have spent thousands of hours studying its pages and have commented on the treatment offered. The net result is that a monstrous errata sheet has been produced which may be found at the back of this manual. In fact, the errata sheet comes in four parts. The first (and the main) portion is a list of errata for all printings. The second part of the errata sheet is an appendix containing errata which could not be simply tabulated but which required some special form of explanation. The third and fourth parts of the errata sheet consist of a list of errata and corresponding appendix which apply only to the first printing (thus if you own a valuable first printing you must use all four portions to correct it). Note that the vast majority of the books available are from the second printing and beyond; for these you should only use the first two parts of the errata sheet.

A difficult choice had to be made regarding the distribution of this solutions manual. The question was whether or not to make it readily available to university students. The concern here was that the availability of these solutions to students while they were using this text in a university course would create a difficulty with homework assignments. Indeed, a professor would perhaps be unwilling to assign homework problems knowing that the solutions were fully documented in a manual available to the students. I thought a long time about this question and wondered whether the availability to students would reduce its value or use in universities. I came to the conclusion that, in fact, it would perhaps increase its usefulness as a university text for a number of reasons. For example, the professor would now be in a position to assign complex problems and still not have to prepare solutions for student use after the assignments were graded. In fact, this would tend to eliminate the grading of assignments, a terrible chore for a university professor. Indeed, if the professor felt the need to assign problems for which no solutions were readily available, then he should be capable of generating variations on the ones offered in the text. Regardless of such reasoning, it should be perfectly clear to anyone familiar with the mischievous mind and enormous energy possessed by graduate students that no matter how hard anyone tried to keep this manual from them, they would surely rise to the challenge and manage to circulate it widely. For example, a solutions manual should, in any case, be made available to those professionals in industry who are using the book for their own purposes; once the book is available to industry, it becomes clear that professionals from industry are often themselves students at a university and that would provide a back door for the solutions manual to find its way to the university. The final conclusion is that the manual is being widely distributed, and I urge you to tell your friends about it if they do not already own a copy.

Now let us discuss the format and the notation used in this solutions manual. Each of the problem statements from the text has been reproduced herein and is immediately followed by its solution. This renders the manual somewhat self-contained and makes it much more convenient to use. At the bottom of each page you will find a problem number (occasionally a range of problem numbers). This may be used as a quick identification as to which problem solution is contained on that page; thus to find a particular problem solution, you need merely flick through the pages while scanning this bottom number. In some cases the problem statement in the text itself required substantive correction (such corrections are included in the errata sheet at the back of the manual); these corrections have already been made in the problem statements in this manual, and any problems which have such substantive corrections are denoted with a dagger (†) next to the problem number itself. In addition, to assist the reader in identifying where the changes have been made in the problem statement, vertical bars (|) have been added alongside the problem statement indicating which lines of text and/or equations have been altered. The exact alteration can be found by careful reference to the errata sheets. (As with any other document, this manual is itself subject to errors and typos. I would welcome any errata you discover in the solutions, in the errata sheet (!) or in the text itself.) You may recall that a solid black rectangle was used in the text to identify equations of importance (these equations then appeared in the Summary of Results in the back of the text); we have used a similar black square to denote additional equations of importance which appear in the solutions manual.

The compilation of the solutions contained herein has been an extremely arduous and difficult task. Many of the students who passed through my queueing theory classes have assisted me greatly in this effort. They are far too numerous to acknowledge individually and so I take this opportunity to acknowledge them as a group. By far the greatest effort was put forward by my student, Richard Gail, and he is listed as a co-author of this manual. I also wish to acknowledge the Advanced Research Projects Agency of the Department of Defense for providing the environment in which this material could be developed.

I probably have no need to offer you the reader the following advice, but I will nevertheless. Surely, you should make every attempt to solve a problem on your own before referring to the solutions contained herein. Queueing theory is a rather difficult subject, but unfortunately it appears deceptively simple to the novice. The only way to gain a true understanding of the material is to work through lots of problems on your own. Only when you choose to check your solution or, in those rare cases, when you run into a serious roadblock, should you use this manual to assist you in your progress. All the details are here — I hope you enjoy the revelations.

LEONARD KLEINROCK

Los Angeles
October 1981

Contents

	PAGE
Solutions for Chapter 2	1
Solutions for Chapter 3	39
Solutions for Chapter 4	69
Solutions for Chapter 5	95
Solutions for Chapter 6	161
Solutions for Chapter 7	171
Solutions for Chapter 8	179
Errata for Volume I	217

Chapter 2

Some Important Random Processes

PROBLEM 2.1.

Consider K independent sources of customers where the interarrival time between customers for each source is exponentially distributed with parameter λ_k (i.e., each source is a Poisson process). Now consider the arrival stream, which is formed by merging the input from each of the K sources defined above. Prove that this merged stream is also Poisson with parameter $\lambda = \lambda_1 + \lambda_2 + \cdots + \lambda_K$.

SOLUTION

Let X_i be the Poisson counting process for the ith source and let $X = X_1 + \cdots + X_K$. The z-transform for X is

$$G(z) = E[z^X] = E[z^{X_1 + \cdots + X_K}]$$

$$= E[z^{X_1}] \cdots E[z^{X_K}] \quad \text{(since the } X_i \text{ are independent)}$$

$$= e^{\lambda_1 t(z-1)} \cdots e^{\lambda_K t(z-1)}$$

$$= e^{(\lambda_1 + \cdots + \lambda_K) t(z-1)}$$

which is the z-transform for a Poisson process with parameter $\lambda_1 + \cdots + \lambda_K$.

PROBLEM 2.2.

Referring back to the previous problem, consider this merged Poisson stream and now assume that we wish to break it up into several branches. Let p_i be the probability that a customer from the merged stream is assigned to the ith substream. If the overall rate is λ customers per second, and if the substream probabilities p_i are chosen for each customer independently, then show that each of these substreams is a Poisson process with rate λp_i.

SOLUTION

Suppose we wish to split the merged Poisson stream into r branches. Let N be the number of arrivals to the merged stream in an interval of duration t, and let N_i for $i = 1, \ldots, r$ be the number assigned to substream i (so that $N = N_1 + \cdots + N_r$). We claim that

$$P[N_1 = n_1, \ldots, N_r = n_r | N = n] = \frac{n!}{n_1! \cdots n_r!} p_1^{n_1} \cdots p_r^{n_r}$$

(the multinomial distribution) as follows. Any specific choice of n_i customers for substream i ($i = 1, \ldots, r$) is easily seen to have probability $p_1^{n_1} \cdots p_r^{n_r}$. The number of such possible choices is equal to the number of ways of picking n_1 out of n, times the number of ways of appropriately assigning $n_2, \ldots, n_r$ from the remaining $n - n_1$. Using induction, this is simply

$$\frac{n!}{(n-n_1)! \, n_1!} \times \frac{(n-n_1)!}{n_2! \cdots n_r!} = \frac{n!}{n_1! \cdots n_r!}$$

which verifies the claim. We finally obtain

$$P[N_1 = n_1, \ldots, N_r = n_r] = \frac{n!}{n_1! \cdots n_r!} p_1^{n_1} \cdots p_r^{n_r} P[N=n]$$

$$= \frac{n!}{n_1! \cdots n_r!} p_1^{n_1} \cdots p_r^{n_r} \frac{(\lambda t)^n}{n!} e^{-\lambda t}$$

$$P[N_1 = n_1, \ldots, N_r = n_r] = \prod_{i=1}^{r} \frac{(\lambda p_i t)^{n_i}}{n_i!} e^{-\lambda p_i t}$$

Thus we have shown that the r substreams form independent Poisson processes, where the ith substream has rate λp_i.

PROBLEM 2.3.

Let $\{X_j\}$ be a sequence of identically distributed mutually independent Bernoulli random variables (with $P[X_j = 1] = p$, and $P[X_j = 0] = 1 - p$). Let $S_N = X_1 + \cdots + X_N$ be the sum of a random number N of the random variables X_j, where N has a Poisson distribution with mean λ. Prove that S_N has a Poisson distribution with mean λp. (In general, the distribution of the sum of a random number of independent random variables is called a compound distribution.)

SOLUTION

Condition on $N = n$. Then the conditional z-transform is
$$G(z|N=n) = E[z^{S_n}] = E[z^{X_1 + \cdots + X_n}] = \left(E[z^{X_1}]\right)^n.$$
But
$$E[z^{X_1}] = \sum_{k=0}^{\infty} P[X_1 = k] z^k = P[X_1 = 0] \cdot 1 + P[X_1 = 1] \cdot z$$

$$= 1 - p + pz$$

$$\therefore \; G(z|N=n) = (1 - p + pz)^n$$

Unconditioning on N, we have
$$G(z) = \sum_{n=0}^{\infty} G(z|N=n) \cdot P[N=n]$$

$$= \sum_{n=0}^{\infty} (1 - p + pz)^n \frac{\lambda^n e^{-\lambda}}{n!}$$

$$= e^{-\lambda} e^{\lambda(1 - p + pz)} = e^{\lambda p (z - 1)}$$

(This we recognize as the z-transform for a Poisson distribution.)

$\therefore \; S_N$ is Poisson with parameter λp.

PROBLEM 2.4.

Find the pdf for the smallest of K independent random variables, each of which is exponentially distributed with parameter λ.

SOLUTION

Let the K random variables be $X_1, X_2, \ldots, X_K$. The random variable of interest is $Y = \min(X_1, X_2, \ldots, X_K)$.

$$P[Y>y] = P[X_1>y, \ldots, X_K>y]$$
$$= P[X_1>y] \cdots P[X_K>y] \quad (X_i \text{ are independent})$$
$$= e^{-\lambda y} \cdots e^{-\lambda y} = e^{-K\lambda y}$$

∴ Y is exponential with parameter $K\lambda$

i.e. $P[Y \leq y] = 1 - e^{-K\lambda y}$. ∎

PROBLEM 2.5.

Consider the homogeneous Markov chain whose state diagram is

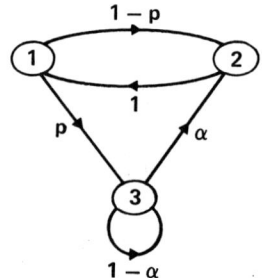

(a) Find **P**, the probability transition matrix.
(b) Under what conditions (if any) will the chain be irreducible and aperiodic?
(c) Solve for the equilibrium probability vector π.
(d) What is the mean recurrence time for state E_2?
(e) For which values of α and p will we have $\pi_1 = \pi_2 = \pi_3$? (Give a physical interpretation of this case.)

SOLUTION

(a)
$$\mathbf{P} = \begin{bmatrix} 0 & 1-p & p \\ 1 & 0 & 0 \\ 0 & \alpha & 1-\alpha \end{bmatrix}$$

(b) Irreducible and aperiodic for all $0 < p \leq 1$ and $0 < \alpha \leq 1$ except $\alpha = p = 1$.

(c) From $\pi = \pi\mathbf{P}$ [$\pi = (\pi_1, \pi_2, \pi_3)$] we obtain only two independent equations, namely

$$\pi_1 = \pi_2$$
$$\pi_2 = (1-p)\pi_1 + \alpha\pi_3$$

Using the conservation of probability, we also have $\pi_1 + \pi_2 + \pi_3 = 1$. Thus

$$\pi_1 = \pi_2 = \frac{\alpha}{p + 2\alpha} \qquad \blacksquare$$

$$\pi_3 = \frac{p}{p + 2\alpha} \qquad \blacksquare$$

(d)
$$\mu_2 = \frac{1}{\pi_2} = \frac{p + 2\alpha}{\alpha} = 2 + \frac{p}{\alpha} \qquad \blacksquare$$

(e) We need

$$\alpha = p \qquad \blacksquare$$

Interpretation: Since each visit to E_1 is followed by exactly one visit to E_2 and vice versa for all p, α we have $\pi_1 = \pi_2$ always. Also $p \, (= \alpha)$ of the time we go from E_1 to E_3 and the average number of steps (or mean time) spent in E_3 per visit is $\frac{1}{1-(1-\alpha)} = \frac{1}{\alpha}$. Thus $\alpha \cdot \frac{1}{\alpha} = 1$ is the average number of visits to E_3 per visit to E_1.

PROBLEM 2.6.

Consider the discrete-state, discrete-time Markov chain whose transition probability matrix is given by

$$\mathbf{P} = \begin{bmatrix} \frac{1}{2} & \frac{1}{2} \\ \frac{3}{4} & \frac{1}{4} \end{bmatrix}$$

(a) Find the stationary state probability vector π.
(b) Find $[\mathbf{I} - z\mathbf{P}]^{-1}$.
(c) Find the general form for $\mathbf{P}^n$.

SOLUTION

(a) $\pi = \pi P$ [$\pi = (\pi_1, \pi_2)$] and $\pi_1 + \pi_2 = 1$. So

$$\pi_1 = \frac{3}{5}, \pi_2 = \frac{2}{5}$$ ∎

(b)

$$I - zP = \begin{bmatrix} 1 - \frac{1}{2}z & -\frac{1}{2}z \\ -\frac{3}{4}z & 1 - \frac{1}{4}z \end{bmatrix}$$

$$\det(I - zP) = \left(1 - \frac{1}{2}z\right)\left(1 - \frac{1}{4}z\right) - \left(-\frac{1}{2}z\right)\left(-\frac{3}{4}z\right)$$

$$= 1 - \frac{3}{4}z - \frac{1}{4}z^2$$

$$= (1-z)\left(1 + \frac{1}{4}z\right)$$

$$(I - zP)^{-1} = \frac{1}{(1-z)\left(1 + \frac{1}{4}z\right)} \begin{bmatrix} 1 - \frac{1}{4}z & \frac{1}{2}z \\ \frac{3}{4}z & 1 - \frac{1}{2}z \end{bmatrix}$$

(c) Using partial fraction expansions,

$$(I - zP)^{-1} = \frac{1}{1-z}\begin{bmatrix} \frac{3}{5} & \frac{2}{5} \\ \frac{3}{5} & \frac{2}{5} \end{bmatrix} + \frac{1}{1 + \frac{1}{4}z}\begin{bmatrix} \frac{2}{5} & -\frac{2}{5} \\ -\frac{3}{5} & \frac{3}{5} \end{bmatrix}$$

Since $(I - zP)^{-1} \Leftrightarrow P^n$ we get

$$P^n = \begin{bmatrix} \frac{3}{5} & \frac{2}{5} \\ \frac{3}{5} & \frac{2}{5} \end{bmatrix} + \left(-\frac{1}{4}\right)^n \begin{bmatrix} \frac{2}{5} & -\frac{2}{5} \\ -\frac{3}{5} & \frac{3}{5} \end{bmatrix}$$ ∎

2.6.

PROBLEM 2.7.

Consider a Markov chain with states $E_0, E_1, E_2, \ldots$ and with transition probabilities

$$p_{ij} = e^{-\lambda} \sum_{n=0}^{j} \binom{i}{n} p^n q^{i-n} \frac{\lambda^{j-n}}{(j-n)!}$$

where $p + q = 1$ $(0 < p < 1)$.
(a) Is this chain irreducible? Periodic? Explain.
(b) We wish to find

$$\pi_i = \text{equilibrium probability of } E_i$$

Write π_i in terms of p_{ij} and π_j for $j = 0, 1, 2, \ldots$.
(c) From (b) find an expression relating $P(z)$ to $P[1 + p(z-1)]$, where

$$P(z) = \sum_{i=0}^{\infty} \pi_i z^i$$

(d) Recursively (i.e., repeatedly) apply the result in (c) to itself and show that the nth recursion gives

$$P(z) = e^{\lambda(z-1)(1 + p + p^2 + \cdots + p^{n-1})} P[1 + p^n(z-1)]$$

(e) From (d) find $P(z)$ and then recognize π_i.

SOLUTION

(a) The given Markov chain is irreducible since $p_{ij} > 0$ for all i, j (since $0 < p < 1$). It is aperiodic since $p_{ii} > 0$ for at least one (and, in fact, all) i.

(b)
$$\pi_i = \sum_{j=0}^{\infty} \pi_j p_{ji}$$

(c)
$$P(z) = \sum_{i=0}^{\infty} \pi_i z^i = \sum_{i=0}^{\infty} \sum_{j=0}^{\infty} \pi_j p_{ji} z^i$$

$$= \sum_{i=0}^{\infty} \sum_{j=0}^{\infty} \pi_j z^i e^{-\lambda} \sum_{n=0}^{i} \binom{j}{n} p^n q^{j-n} \frac{\lambda^{i-n}}{(i-n)!}$$

Since $\sum_{i=0}^{\infty} \sum_{n=0}^{i} = \sum_{n=0}^{\infty} \sum_{i=n}^{\infty}$ we have

2.7.

$$P(z) = e^{-\lambda}\sum_{j=0}^{\infty}\pi_j \sum_{n=0}^{\infty}\binom{j}{n}p^n q^{j-n}\sum_{i=n}^{\infty}z^i \frac{\lambda^{i-n}}{(i-n)!}$$

$$= e^{-\lambda}e^{\lambda z}\sum_{j=0}^{\infty}\pi_j \sum_{n=0}^{\infty}\binom{j}{n}(pz)^n q^{j-n}$$

Since $\binom{j}{n}=0$ for $n>j$, then

$$P(z) = e^{\lambda(z-1)}\sum_{j=0}^{\infty}\pi_j(pz+q)^j$$

$$\therefore P(z) = e^{\lambda(z-1)}P[1+p(z-1)] \qquad \blacksquare$$

(d) The expression is clearly true for $n=1$ by part (c). By induction, assume it is true for arbitrary n, in which case

$$P(z) = e^{\lambda(z-1)(1+p+p^2+\cdots+p^{n-1})}P[1+p^n(z-1)]$$

Now substitute $1+p^n(z-1)$ for z in the result from part (c) to give

$$P[1+p^n(z-1)] = e^{\lambda p^n(z-1)}P[1+p^{n+1}(z-1)]$$

Thus

$$P(z) = e^{\lambda(z-1)(1+p+p^2+\cdots+p^n)}P[1+p^{n+1}(z-1)]$$

which completes the induction.

(e) From (d),

$$P(z) = e^{\lambda(z-1)(1+p+p^2+\cdots+p^n)}P[1+p^{n+1}(z-1)]$$

for all $n=0,1,2,\ldots$. Consider this expression as $n\to\infty$. This gives

$$P(z) = e^{\lambda(z-1)\sum_{n=0}^{\infty}p^n}P(1) \quad (p^{n+1}\to 0 \text{ since } 0<p<1)$$

or

$$P(z) = e^{\lambda(z-1)\frac{1}{1-p}} \quad (P(1)=1).$$

From Eq. (2.134) we recognize this as the z-transform of a Poisson distribution with parameter $\frac{\lambda}{q}$. Hence

$$\pi_i = \frac{\left(\frac{\lambda}{q}\right)^i e^{-\frac{\lambda}{q}}}{i!} \qquad \blacksquare$$

2.7.

PROBLEM 2.8.

Show that any point in or on the equilateral triangle of unit height shown in Figure 2.6 represents a three-component probability vector in the sense that the sum of the distances from any such point to each of the three sides must always equal unity.

SOLUTION

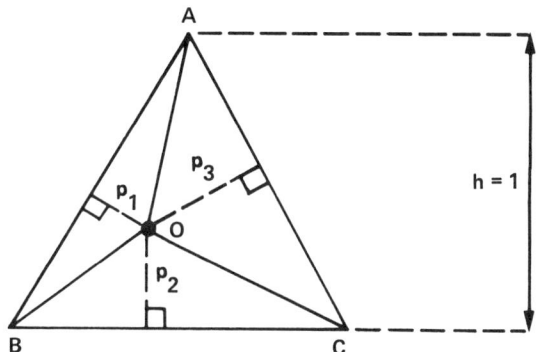

Let $\triangle ABC$ denote triangle ABC and let $|AB|$ denote the length of side AB. Then

$$\text{Area } \triangle ABC = \text{Area } \triangle AOB + \text{Area } \triangle BOC + \text{Area } \triangle AOC$$

$$\therefore |BC|\frac{h}{2} = |AB|\frac{p_1}{2} + |BC|\frac{p_2}{2} + |AC|\frac{p_3}{2}$$

But $|BC| = |AB| = |AC|$ since $\triangle ABC$ is equilateral. So

$$h = p_1 + p_2 + p_3$$

Since $h = 1$ we have

$$p_1 + p_2 + p_3 = 1.$$

2.8.

PROBLEM 2.9.

Consider a pure birth process with constant birth rate λ. Let us consider an interval of length T, which we divide up into m segments each of length T/m. Define $\Delta t = T/m$.
(a) For Δt small, find the probability that a single arrival occurs in each of exactly k of the m intervals and that no arrivals occur in the remaining $m - k$ intervals.
(b) Consider the limit as $\Delta t \to 0$, that is, as $m \to \infty$ for fixed T, and evaluate the probability $P_k(T)$ that exactly k arrivals occur in the interval of length T.

SOLUTION

$$P[1 \text{ arrival in } \Delta t] = \lambda \Delta t + o(\Delta t)$$
$$P[0 \text{ arrivals in } \Delta t] = 1 - \lambda \Delta t + o(\Delta t)$$

(a)

$$P\begin{bmatrix} 1 \text{ arrival in exactly } k \text{ of the } m \text{ intervals} \\ \text{and } 0 \text{ arrivals in the remaining } m - k \text{ intervals} \end{bmatrix} =$$

$$\begin{Bmatrix} \text{number of ways} \\ k \text{ cells may be} \\ \text{selected from } m \end{Bmatrix} [\lambda \Delta t + o(\Delta t)]^k [1 - \lambda \Delta t + o(\Delta t)]^{m-k}$$

$$= \binom{m}{k} (\lambda \Delta t)^k (1 - \lambda \Delta t)^{m-k} + o(\Delta t)$$

(b) As $\Delta t \to 0$, since $\Delta t = T/m$, then $m \to \infty$ (and $o(\Delta t) \to 0$).

$$P_k(T) = \lim_{\Delta t \to 0} \left[\binom{m}{k} (\lambda \Delta t)^k (1 - \lambda \Delta t)^{m-k} + o(\Delta t) \right]$$

$$= \lim_{m \to \infty} \frac{m!}{k!(m-k)!} \left(\frac{\lambda T}{m} \right)^k \left(1 - \frac{\lambda T}{m} \right)^{m-k}$$

$$= \frac{(\lambda T)^k}{k!} \lim_{m \to \infty} \left(\frac{m}{m} \right) \left(\frac{m-1}{m} \right) \cdots \left(\frac{m-k+1}{m} \right) \left(1 - \frac{\lambda T}{m} \right)^{m-k}$$

$$= \frac{(\lambda T)^k}{k!} \lim_{m \to \infty} \left(1 - \frac{\lambda T}{m} \right)^{m-k} \quad (k \text{ fixed})$$

$$\therefore P_k(T) = \frac{(\lambda T)^k}{k!} e^{-\lambda T} \qquad \blacksquare$$

2.9.

PROBLEM 2.10.

Consider a population of bacteria of size $N(t)$ at time t for which $N(0) = 1$. We consider this to be a pure birth process in which any member of the population will split into two new members in the interval $(t, t+\Delta t)$ with probability $\lambda \Delta t + o(\Delta t)$ or will remain unchanged in this interval with probability $1 - \lambda \Delta t + o(\Delta t)$ as $\Delta t \to 0$.

(a) Let $P_k(t) = P[N(t) = k]$ and write down the set of differential-difference equations that must be satisfied by these probabilities.

(b) From part (a) show that the z-transform $P(z, t)$ for $N(t)$ must satisfy
$$P(z, t) = \frac{ze^{-\lambda t}}{1 - z + ze^{-\lambda t}}$$

(c) Find $E[N(t)]$.

(d) Solve for $P_k(t)$.

(e) Solve for $P(z, t)$, $E[N(t)]$ and $P_k(t)$ that satisfy the initial condition $N(0) = n \geq 1$.

(f) Consider the corresponding deterministic problem in which each bacterium splits into two every $1/\lambda$ sec and compare with the answer in part (c).

SOLUTION

(a)

$$P_k(t+\Delta t) = (1 - k\lambda \Delta t) P_k(t) + (k-1)\lambda \Delta t \, P_{k-1}(t) + o(\Delta t)$$

$$P_k(t+\Delta t) - P_k(t) = (k-1)\lambda \Delta t \, P_{k-1}(t) - k\lambda \Delta t \, P_k(t) + o(\Delta t)$$

$$\frac{dP_k(t)}{dt} = (k-1)\lambda P_{k-1}(t) - k\lambda P_k(t) \quad k \geq 1 \qquad \blacksquare$$

$$\frac{dP_0(t)}{dt} = 0 \text{ since } P_0(t) = 0 \text{ for all } t \qquad \blacksquare$$

$$(P_1(0) = 1 \text{ and pure birth})$$

(b) First we must form the z-transform of the above difference equations. Let $P(z, t) = \sum_{k=0}^{\infty} P_k(t) z^k$. We observe that $\dfrac{\partial P(z, t)}{\partial t} = \sum_{k=0}^{\infty} \dfrac{dP_k(t)}{dt} z^k$.
Using part (a), multiply the kth equation by z^k and sum to yield

$$\frac{\partial P(z,t)}{\partial t} = \sum_{k=1}^{\infty}\left[(k-1)\lambda P_{k-1}(t)z^k - k\lambda P_k(t)z^k\right]$$

$$= \lambda z^2 \sum_{k=2}^{\infty}(k-1)P_{k-1}(t)z^{k-2} - \lambda z \sum_{k=1}^{\infty} kP_k(t)z^{k-1}$$

$$= \lambda z^2 \frac{\partial P(z,t)}{\partial z} - \lambda z \frac{\partial P(z,t)}{\partial z}$$

Thus

$$\frac{\partial P(z,t)}{\partial t} = \lambda z(z-1)\frac{\partial P(z,t)}{\partial z} \qquad \blacksquare$$

or

$$\frac{\partial P(z,t)}{\partial t} + \lambda z(1-z)\frac{\partial P(z,t)}{\partial z} = 0 \qquad \blacksquare$$

Now we must solve this partial differential equation (p.d.e.). Anticipating the solution to Exercise 2.14(b), let us solve the (more general) p.d.e.

$$\frac{\partial P(z,t)}{\partial t} + (\lambda z - \gamma)(1-z)\frac{\partial P(z,t)}{\partial z} = 0$$

[Note that the equation derived above is the special case when $\gamma = 0$.] We solve this linear first order p.d.e. by the Lagrange method as described on pages 696-697 of [SYSK 60] (see also *Fundamentals of Queueing Theory* by Gross and Harris, pp 115-116 (Wiley 1974)). Thus we first must solve the subsidiary equation

$$\frac{dt}{1} = \frac{dz}{(\lambda z - \gamma)(1-z)}$$

which becomes, upon using partial fraction expansion,

$$\frac{dt}{1} = \frac{dz}{\lambda - \gamma}\left[\frac{\lambda}{\lambda z - \gamma} + \frac{1}{1-z}\right]$$

or

$$\frac{\lambda \, dz}{\lambda z - \gamma} - \frac{dz}{z-1} = (\lambda - \gamma)\, dt$$

Integrating yields

$$\log_e\left(\frac{\lambda z - \gamma}{z-1}\right) = (\lambda - \gamma)t + C$$

and thus

$$\frac{\lambda z - \gamma}{z - 1} = K\, e^{(\lambda - \gamma)t}$$

2.10.

or
$$\frac{\gamma - \lambda z}{1-z} e^{-(\lambda-\gamma)t} = K$$

The general solution is now given as
$$P(z,t) = f\left[\frac{\gamma - \lambda z}{1-z} e^{-(\lambda-\gamma)t}\right]$$

where the function f must be determined by the initial condition $P(z,0) = z$. This condition gives
$$f\left[\frac{\gamma - \lambda z}{1-z}\right] = z$$

and so f satisfies
$$f(y) = \frac{y - \gamma}{y - \lambda}$$

Therefore
$$P(z,t) = \frac{\dfrac{\gamma - \lambda z}{1-z} e^{-(\lambda-\gamma)t} - \gamma}{\dfrac{\gamma - \lambda z}{1-z} e^{-(\lambda-\gamma)t} - \lambda}$$

$$= \frac{\gamma - \lambda z - \gamma(1-z) e^{(\lambda-\gamma)t}}{\gamma - \lambda z - \lambda(1-z) e^{(\lambda-\gamma)t}}$$

$$P(z,t) = \frac{\gamma(1 - e^{(\lambda-\gamma)t}) - (\lambda - \gamma e^{(\lambda-\gamma)t})z}{\gamma - \lambda e^{(\lambda-\gamma)t} - \lambda(1 - e^{(\lambda-\gamma)t})z} \qquad\blacksquare$$

which is the solution of the (more general) p.d.e. The special case $\gamma = 0$ gives the desired solution
$$P(z,t) = \frac{ze^{-\lambda t}}{1 - z + ze^{-\lambda t}} \qquad\blacksquare$$

(c)
$$E[N(t)] = \left.\frac{\partial P(z,t)}{\partial z}\right|_{z=1} = \left.\frac{e^{-\lambda t}}{[1 - z + ze^{-\lambda t}]^2}\right|_{z=1}$$

$$E[N(t)] = e^{\lambda t} \qquad\blacksquare$$

(d) We have $P(z,t) = \dfrac{ze^{-\lambda t}}{1 - [1 - e^{-\lambda t}]z}$. To invert this z-transform we note from Entry 6 Table I.2 that

2.10.

$$\frac{e^{-\lambda t}}{1-[1-e^{-\lambda t}]z} \Leftrightarrow e^{-\lambda t}[1-e^{-\lambda t}]^k$$

Using this result and Entry 7 Table I.1 we have

$$P_k(t) = \begin{cases} 0 & k=0 \text{ (pure birth)} \\ e^{-\lambda t}[1-e^{-\lambda t}]^{k-1} & k \geq 1 \end{cases}$$ ■

(e) $N(0) = n \geq 1$. This is the same as having n independent birth processes each with $N(0) = 1$. The z-transform is the n-fold product of the z-transform derived in part (b),

$$\text{i.e. } P(z,t) = \left(\frac{ze^{-\lambda t}}{1-z+ze^{-\lambda t}}\right)^n.$$ ■

Thus

$$E[N(t)] = ne^{\lambda t} \quad \text{(sum of expectations).}$$ ■

$P(z,t) = \dfrac{e^{-n\lambda t}z^n}{\left(1-[1-e^{-\lambda t}]z\right)^n}$; inverting this we first find from Entry 13 Table I.2 that

$$\frac{e^{-n\lambda t}}{\left(1-[1-e^{-\lambda t}]z\right)^n} \Leftrightarrow e^{-n\lambda t}\binom{k+n-1}{n-1}[1-e^{-\lambda t}]^k$$

Now applying Entry 8 Table I.1 we have

$$P_k(t) = \begin{cases} 0 & k<n \text{ (pure birth)} \\ \binom{k-1}{n-1}e^{-n\lambda t}[1-e^{-\lambda t}]^{k-n} & k \geq n \end{cases}$$ ■

(f) In the deterministic problem, each individual will split into 2 every $1/\lambda$ seconds exactly. Since $N(0) = 1$, and since every $1/\lambda$ seconds the population doubles, then

$$N(t) = 2^{\lambda t}, \quad t = \frac{k}{\lambda}, \quad k = 0, 1, 2, \ldots$$

Note that this growth rate is less than $e^{\lambda t}$ for the random case. This is exactly analogous to the situation when $\$1.00$ is compounded annually at 100% interest to yield $\$2^n$ in n years as opposed to $\$e^n$ if compounded continuously.

2.10.

PROBLEM 2.11.

Consider a birth-death process with coefficients

$$\lambda_k = \begin{cases} \lambda & k=0 \\ 0 & k \neq 0 \end{cases} \quad \mu_k = \begin{cases} \mu & k=1 \\ 0 & k \neq 1 \end{cases}$$

which corresponds to an M/M/1 queueing system where there is no room for waiting customers.
(a) Give the differential-difference equations for $P_k(t)$ ($k = 0, 1$).
(b) Solve these equations and express the answers in terms of $P_0(0)$ and $P_1(0)$.

SOLUTION

(a)
$$\frac{dP_0(t)}{dt} = -\lambda P_0(t) + \mu P_1(t)$$

$$\frac{dP_1(t)}{dt} = -\mu P_1(t) + \lambda P_0(t)$$

(b) Since $P_0(t) + P_1(t) = 1$, we may rewrite the first equation from part (a) as

$$\frac{dP_0(t)}{dt} + (\lambda + \mu) P_0(t) = \mu$$

As in Section I.4, we recognize that the homogeneous solution must be of the form $Ae^{-(\lambda+\mu)t}$; we find that the particular solution is a constant (say B) and so

$$P_0(t) = B + Ae^{-(\lambda+\mu)t}.$$

Substituting into our differential equation, we have $B = \dfrac{\mu}{\lambda + \mu}$. Evaluating the solution at $t = 0$, we have $A = P_0(0) - \dfrac{\mu}{\lambda + \mu}$. Hence

$$P_0(t) = \frac{\mu}{\lambda + \mu} + \left(P_0(0) - \frac{\mu}{\lambda + \mu}\right) e^{-(\lambda+\mu)t}.$$

$P_1(t)$ may be found from $P_1(t) = 1 - P_0(t)$ or by symmetry of the defining equations from part (a) to yield

$$P_1(t) = \frac{\lambda}{\lambda+\mu} + \left[P_1(0) - \frac{\lambda}{\lambda+\mu}\right]e^{-(\lambda+\mu)t}. \qquad \blacksquare$$

PROBLEM 2.12.

Consider a birth-death queueing system in which
$$\lambda_k = \lambda \qquad k \geq 0$$
$$\mu_k = k\mu \qquad k \geq 0$$

(a) For all k, find the differential-difference equations for
$$P_k(t) = P[k \text{ in system at time } t]$$

(b) Define the z-transform
$$P(z, t) = \sum_{k=0}^{\infty} P_k(t) z^k$$
and find the *partial* differential equation that $P(z, t)$ must satisfy.

(c) Show that the solution to this equation is
$$P(z, t) = \exp\left[\frac{\lambda}{\mu}(1 - e^{-\mu t})(z - 1)\right]$$
with the initial condition $P_0(0) = 1$.

(d) Comparing the solution in part (c) with Eq. (2.134), give the expression for $P_k(t)$ by inspection.

(e) Find the limiting values for these probabilities as $t \to \infty$.

SOLUTION

(a)
$$\frac{dP_k(t)}{dt} = -(\lambda + k\mu)P_k(t) + \lambda P_{k-1}(t) + (k+1)\mu P_{k+1}(t) \qquad k \geq 1 \qquad \blacksquare$$

$$\frac{dP_0(t)}{dt} = -\lambda P_0(t) + \mu P_1(t) \qquad k = 0 \qquad \blacksquare$$

(b) For $k \geq 1$, we multiply the kth equation by z^k, sum on k, and then add the equation for $k = 0$ to get

$$\sum_{k=0}^{\infty} \frac{d P_k(t)}{dt} z^k = -\lambda \sum_{k=0}^{\infty} P_k(t) z^k - \mu z \sum_{k=1}^{\infty} k P_k(t) z^{k-1}$$

$$+ \lambda z \sum_{k=1}^{\infty} P_{k-1}(t) z^{k-1} + \mu \sum_{k=0}^{\infty} (k+1) P_{k+1}(t) z^k$$

$$\frac{\partial P(z,t)}{\partial t} = -\lambda P(z,t) - \mu z \frac{\partial P(z,t)}{\partial z} + \lambda z P(z,t) + \mu \frac{\partial P(z,t)}{\partial z}$$

$$\frac{\partial P(z,t)}{\partial t} = \lambda(z-1) P(z,t) - \mu(z-1) \frac{\partial P(z,t)}{\partial z} \qquad \blacksquare$$

(c) The partial differential equation derived in part (b) may be solved by the method used in Exercise 2.10(b) (see the references mentioned there). Writing our equation in the form

$$\frac{\partial P(z,t)}{\partial t} + \mu(z-1) \frac{\partial P(z,t)}{\partial z} = \lambda(z-1) P(z,t)$$

we see that the following subsidiary equations must first be solved

$$\frac{dt}{1} = \frac{dz}{\mu(z-1)} = \frac{d P(z,t)}{\lambda(z-1) P(z,t)}$$

The first equation becomes

$$\frac{dz}{z-1} = \mu \, dt$$

which yields, upon integration,

$$\log_e(z-1) = \mu t + C_1$$

and thus

$$z - 1 = K_1 e^{\mu t}$$

or

$$(z-1) e^{-\mu t} = K_1$$

The second equation becomes

$$\frac{d P(z,t)}{P(z,t)} = \frac{\lambda}{\mu} dz$$

which yields

$$\log_e P(z,t) = \frac{\lambda}{\mu} z + C_2$$

2.12.

and thus
$$P(z,t) = K_2 e^{\frac{\lambda}{\mu}z}$$
or
$$P(z,t) e^{-\frac{\lambda}{\mu}z} = K_2$$

The general solution is now given as
$$P(z,t) e^{-\frac{\lambda}{\mu}z} = f[(z-1)e^{-\mu t}]$$
where the function f must be determined by the initial condition $P_0(0) = 1$. This condition (which is equivalent to $P(z,0) = 1$) gives
$$e^{-\frac{\lambda}{\mu}z} = f(z-1)$$
and so f satisfies
$$f(y) = e^{-\frac{\lambda}{\mu}(y+1)}$$
Therefore
$$P(z,t) = f[(z-1)e^{-\mu t}]e^{\frac{\lambda}{\mu}z} = e^{-\frac{\lambda}{\mu}[(z-1)e^{-\mu t}+1]}e^{\frac{\lambda}{\mu}z}$$
which gives the desired solution
$$P(z,t) = \exp\left[\frac{\lambda}{\mu}(1-e^{-\mu t})(z-1)\right]$$

(The interested reader may wish to substitute this solution into the p.d.e. above.)

(d) $P_k(t)$ is clearly Poisson with parameter $\frac{\lambda}{\mu}[1-e^{-\mu t}]$.
$$P_k(t) = \frac{\left(\frac{\lambda}{\mu}\right)^k [1-e^{-\mu t}]^k e^{-\frac{\lambda}{\mu}[1-e^{-\mu t}]}}{k!} \quad \blacksquare$$

(e) As $t \to \infty$, $e^{-\mu t} \to 0$ and $P_k(t) \to p_k$.
$$p_k = \frac{\left(\frac{\lambda}{\mu}\right)^k e^{-\frac{\lambda}{\mu}}}{k!} \quad \left(\text{Poisson with parameter } \frac{\lambda}{\mu}\right) \quad \blacksquare$$

2.12.

PROBLEM 2.13.

Consider a system in which the birth rate decreases and the death rate increases as the number in the system k increases, that is,

$$\lambda_k = \begin{cases} (K-k)\lambda & k \leqslant K \\ 0 & k \geqslant K \end{cases} \quad \mu_k = \begin{cases} k\mu & k \leqslant K \\ 0 & k \geqslant K \end{cases}$$

Write down the differential-difference equations for

$$P_k(t) = P[k \text{ in system at time } t].$$

SOLUTION

$$\frac{dP_0(t)}{dt} = \mu P_1(t) - K\lambda P_0(t) \quad k=0 \quad \blacksquare$$

$$\frac{dP_k(t)}{dt} = \lambda(K-k+1)P_{k-1}(t) + (k+1)\mu P_{k+1}(t)$$

$$- [(K-k)\lambda + k\mu]P_k(t) \quad k=1, 2, \ldots, K-1 \quad \blacksquare$$

$$\frac{dP_K(t)}{dt} = \lambda P_{K-1}(t) - K\mu P_K(t) \quad k=K \quad \blacksquare$$

PROBLEM 2.14.

Consider the case of a linear birth-death process in which $\lambda_k = k\lambda$ and $\mu_k = k\mu$.
(a) Find the partial-differential equation that must be satisfied by $P(z, t)$ as defined in Eq. (2.153).
(b) Assuming that the population size is one at time zero, show that the function that satisfies the equation in part (a) is

$$P(z, t) = \frac{\mu(1 - e^{(\lambda-\mu)t}) - (\lambda - \mu e^{(\lambda-\mu)t})z}{\mu - \lambda e^{(\lambda-\mu)t} - \lambda(1 - e^{(\lambda-\mu)t})z}$$

(c) Expanding $P(z, t)$ in a power series show that
$$P_k(t) = [1 - \alpha(t)][1 - \beta(t)][\beta(t)]^{k-1} \quad k = 1, 2, \ldots$$
$$P_0(t) = \alpha(t)$$
and find $\alpha(t)$ and $\beta(t)$.
(d) Find the mean and variance for the number in system at time t.
(e) Find the limiting probability that the population dies out by time t for $t \to \infty$.

SOLUTION

(a)
$$\frac{dP_k(t)}{dt} = -(k\lambda + k\mu)P_k(t) + (k-1)\lambda P_{k-1}(t) + (k+1)\mu P_{k+1}(t) \quad k \geq 1$$

$$\frac{dP_0(t)}{dt} = \mu P_1(t) \quad k = 0$$

Multiplying the kth equation by z^k and summing we obtain

$$\sum_{k=0}^{\infty} \frac{dP_k(t)}{dt} z^k = -(\lambda + \mu) \sum_{k=1}^{\infty} kP_k(t) z^k + \lambda \sum_{k=1}^{\infty} (k-1) P_{k-1}(t) z^k$$

$$+ \mu \sum_{k=0}^{\infty} (k+1) P_{k+1}(t) z^k$$

$$\frac{\partial P(z, t)}{\partial t} = -(\lambda + \mu) z \frac{\partial P(z, t)}{\partial z} + \lambda z^2 \frac{\partial P(z, t)}{\partial z} + \mu \frac{\partial P(z, t)}{\partial z}$$

$$\frac{\partial P(z, t)}{\partial t} = (\lambda z - \mu)(z - 1) \frac{\partial P(z, t)}{\partial z} \quad \blacksquare$$

(b) As part of the solution to Exercise 2.10(b), the equation
$$\frac{\partial P(z, t)}{\partial t} + (\lambda z - \gamma)(1 - z) \frac{\partial P(z, t)}{\partial z} = 0$$
was solved. We see that the partial differential equation derived in part (a) above is the same as this equation with μ in place of γ. Thus the solution obtained in Exercise 2.10(b) may be used, and we find that

$$P(z, t) = \frac{\mu(1 - e^{(\lambda - \mu)t}) - (\lambda - \mu e^{(\lambda - \mu)t}) z}{\mu - \lambda e^{(\lambda - \mu)t} - \lambda(1 - e^{(\lambda - \mu)t}) z} \quad \blacksquare$$

as desired.

2.14.

(c) Set

$$a = a(t) = \mu(1 - e^{(\lambda-\mu)t}) \qquad b = b(t) = \lambda - \mu e^{(\lambda-\mu)t}$$
$$c = c(t) = \mu - \lambda e^{(\lambda-\mu)t} \qquad d = d(t) = \lambda(1 - e^{(\lambda-\mu)t})$$

Then

$$P(z, t) = \frac{a - bz}{c - dz} = \frac{b}{d} + \frac{ad - bc}{cd\left(1 - \frac{d}{c}z\right)}$$

$$P(z, t) = \frac{b}{d} + \frac{ad - bc}{cd} \sum_{k=0}^{\infty} \left(\frac{d}{c}z\right)^k$$

$k = 0$: $\quad P_0(t) = P(0, t) = \dfrac{b}{d} + \dfrac{ad - bc}{cd} = \dfrac{a}{c}$

$$P_0(t) = \frac{a}{c} = \frac{\mu(1 - e^{(\lambda-\mu)t})}{\mu - \lambda e^{(\lambda-\mu)t}} \triangleq \alpha(t)$$

$k \geq 1$: $\quad P_k(t) = \dfrac{ad - bc}{cd}\left(\dfrac{d}{c}\right)^k = \dfrac{ad - bc}{cd}\left(\dfrac{d}{c}\right)\left(\dfrac{d}{c}\right)^{k-1}$

$$P_k(t) = \frac{ad - bc}{c^2}\left(\frac{d}{c}\right)^{k-1}$$

We take advantage of the fact that $a - b = c - d = \mu - \lambda$. Thus

$$ad - bc = ad + (c - a - d)c = (c - a)(c - d)$$

and so

$$P_k(t) = \frac{(c-a)(c-d)}{c^2}\left(\frac{d}{c}\right)^{k-1} = \left(1 - \frac{a}{c}\right)\left(1 - \frac{d}{c}\right)\left(\frac{d}{c}\right)^{k-1}$$

Therefore

$$P_k(t) = [1 - \alpha(t)][1 - \beta(t)][\beta(t)]^{k-1}$$

where

$$\alpha(t) \triangleq \frac{a}{c} = \frac{\mu(1 - e^{(\lambda-\mu)t})}{\mu - \lambda e^{(\lambda-\mu)t}} \qquad \blacksquare$$

and

$$\beta(t) \triangleq \frac{d}{c} = \frac{\lambda(1 - e^{(\lambda-\mu)t})}{\mu - \lambda e^{(\lambda-\mu)t}} \qquad \blacksquare$$

2.14.

(d)
$$\bar{N}(t) = \left.\frac{\partial P(z,t)}{\partial z}\right|_{z=1}$$

Recall that
$$P(z,t) = \frac{a-bz}{c-dz}$$

and so
$$\frac{\partial P(z,t)}{\partial z} = \frac{ad-bc}{(c-dz)^2} = \frac{(c-a)(c-d)}{(c-dz)^2}$$

Evaluating this partial derivative at $z = 1$, we find
$$\bar{N}(t) = e^{(\lambda-\mu)t}. \qquad \blacksquare$$

Now
$$\sigma^2_{N(t)} = \overline{N^2(t)} - \left[\bar{N}(t)\right]^2$$
$$= \sum_{k=1}^{\infty} k(k-1) P_k(t) + \bar{N}(t) - \left[\bar{N}(t)\right]^2$$

But
$$\sum_{k=1}^{\infty} k(k-1) P_k(t) = \left.\frac{\partial^2 P(z,t)}{\partial z^2}\right|_{z=1} = \left.\frac{2d(c-a)(c-d)}{(c-dz)^3}\right|_{z=1}$$

and so
$$\sum_{k=1}^{\infty} k(k-1) P_k(t) = \frac{2\lambda e^{(\lambda-\mu)t}}{\mu-\lambda}(1 - e^{(\lambda-\mu)t})$$

$$\therefore \sigma^2_{N(t)} = \frac{\mu+\lambda}{\mu-\lambda} e^{(\lambda-\mu)t}(1 - e^{(\lambda-\mu)t}) \qquad \blacksquare$$

(e)
$$p_0 = \lim_{t\to\infty} P_0(t) = \lim_{t\to\infty} \frac{\mu(1-e^{(\lambda-\mu)t})}{\mu - \lambda e^{(\lambda-\mu)t}}$$

$$p_0 = \begin{cases} 1 & \text{if } \lambda \leq \mu \\ \dfrac{\mu}{\lambda} & \text{if } \lambda > \mu \end{cases} \qquad \blacksquare$$

2.14.

PROBLEM 2.15.

Consider a linear birth-death process for which $\lambda_k = k\lambda + \alpha$ and $\mu_k = k\mu$.
(a) Find the differential-difference equations that must be satisfied by $P_k(t)$.
(b) From (a) find the partial-differential equation that must be satisfied by the time-dependent transform defined as

$$P(z, t) = \sum_{k=0}^{\infty} P_k(t) z^k$$

(c) What is the value of $P(1, t)$? Give a verbal interpretation for the expression

$$\overline{N}(t) = \lim_{z \to 1} \frac{\partial}{\partial z} P(z, t)$$

(d) Assuming that the population size begins with i members at time 0, find an ordinary differential equation for $\overline{N}(t)$ and then solve for $\overline{N}(t)$. Consider the case $\lambda = \mu$ as well as $\lambda \neq \mu$.
(e) Find the limiting value for $\overline{N}(t)$ in the case $\lambda < \mu$ (as $t \to \infty$).

SOLUTION

(a)

$$\frac{dP_k(t)}{dt} = -[k(\mu + \lambda) + \alpha] P_k(t) + [(k-1)\lambda + \alpha] P_{k-1}(t)$$

$$+ (k+1)\mu P_{k+1}(t) \quad k \geq 1 \quad \blacksquare$$

$$\frac{dP_0(t)}{dt} = \mu P_1(t) - \alpha P_0(t) \quad k = 0 \quad \blacksquare$$

(b)

$$\sum_{k=0}^{\infty} \frac{dP_k(t)}{dt} z^k = -(\lambda + \mu) \sum_{k=1}^{\infty} k P_k(t) z^k - \alpha \sum_{k=0}^{\infty} P_k(t) z^k$$

$$+ \lambda \sum_{k=1}^{\infty} (k-1) P_{k-1}(t) z^k + \alpha \sum_{k=1}^{\infty} P_{k-1}(t) z^k + \mu \sum_{k=0}^{\infty} (k+1) P_{k+1}(t) z^k$$

2.15.

$$\frac{\partial P(z,t)}{\partial t} = -(\lambda+\mu)z\frac{\partial P(z,t)}{\partial z} - \alpha P(z,t) + \lambda z^2 \frac{\partial P(z,t)}{\partial z}$$

$$+ \alpha z P(z,t) + \mu \frac{\partial P(z,t)}{\partial z}$$

$$\frac{\partial P(z,t)}{\partial t} = (z-1)\left[(\lambda z - \mu)\frac{\partial P(z,t)}{\partial z} + \alpha P(z,t)\right] \qquad \blacksquare$$

(c)

$$P(1,t) = \sum_{k=0}^{\infty} P_k(t) z^k \bigg|_{z=1} = 1 \quad \text{for all } t$$

$$\overline{N}(t) = \lim_{z \to 1} \frac{\partial P(z,t)}{\partial z} = \lim_{z \to 1} \frac{\partial}{\partial z} \sum_{k=0}^{\infty} P_k(t) z^k$$

$$= \lim_{z \to 1} \sum_{k=1}^{\infty} k P_k(t) z^{k-1} = \sum_{k=1}^{\infty} k P_k(t)$$

Thus $\overline{N}(t)$ = average number in system at time t.

(d) We must form the limit as $z \to 1$ in the partial differential equation developed in part (b); this will give us the required equation for $\overline{N}(t)$. To this end, we must evaluate

$$\lim_{z \to 1} \frac{\partial P(z,t)}{\partial t} \cdot \frac{1}{z-1} = \lim_{z \to 1} \frac{\partial^2 P(z,t)}{\partial z \partial t} \cdot 1 \quad \text{(by L'Hospital)}$$

$$= \frac{d}{dt} \lim_{z \to 1} \frac{\partial P(z,t)}{\partial z} = \frac{d}{dt} \overline{N}(t)$$

and also

$$\lim_{z \to 1} \left[(\lambda z - \mu)\frac{\partial P(z,t)}{\partial z} + \alpha P(z,t)\right] = (\lambda - \mu)\overline{N}(t) + \alpha$$

Thus

$$\frac{d}{dt} \overline{N}(t) = (\lambda - \mu)\overline{N}(t) + \alpha \qquad \blacksquare$$

(i) $\lambda = \mu$

$\frac{d}{dt} \overline{N}(t) = \alpha$ and thus

$$\overline{N}(t) = \alpha t + \overline{N}(0) = \alpha t + i \qquad \blacksquare$$

2.15.

(ii) $\lambda \neq \mu$

$\frac{d}{dt}\bar{N}(t) - (\lambda - \mu)\bar{N}(t) = \alpha$. Solving this differential equation as in part (b) of Exercise 2.11, we obtain

$$\bar{N}(t) = \frac{\alpha}{\mu - \lambda} + \left(i - \frac{\alpha}{\mu - \lambda}\right)e^{(\lambda - \mu)t}$$ ∎

(e) For the case $\lambda < \mu$, as $t \to \infty$ then $e^{(\lambda - \mu)t} \to 0$. Hence

$$\lim_{t \to \infty} \bar{N}(t) = \frac{\alpha}{\mu - \lambda} \quad \text{for } \lambda < \mu$$ ∎

PROBLEM 2.16.

Consider the equations of motion in Eq. (2.148) and define the Laplace transform

$$P_k^*(s) = \int_0^\infty P_k(t) e^{-st}\, dt$$

For our initial condition we will assume $P_0(t) = 1$ for $t = 0$. Transform Eq. (2.148) to obtain a set of linear difference equations in $\{P_k^*(s)\}$.

(a) Show that the solution to the set of equations is

$$P_k^*(s) = \frac{\prod_{i=0}^{k-1} \lambda_i}{\prod_{i=0}^{k}(s + \lambda_i)}$$

(b) From (a) find $P_k(t)$ for the case $\lambda_i = \lambda$ ($i = 0, 1, 2, \ldots$).

SOLUTION

Transforming the equations of motion, we get

$$sP_k^*(s) - P_k(0) = -\lambda_k P_k^*(s) + \lambda_{k-1} P_k^*(s) \quad k \geq 1$$

and

$$sP_0^*(s) - P_0(0) = -\lambda_0 P_0^*(s) \quad k = 0$$

The initial condition $P_0(0) = 1$ implies $P_k(0) = 0$ for $k \geq 1$. Thus
$$(s+\lambda_k)P_k^*(s) = \lambda_{k-1}P_{k-1}^*(s) \quad k \geq 1 \qquad \blacksquare$$
and
$$(s+\lambda_0)P_0^*(s) = 1 \quad k=0 \qquad \blacksquare$$

(a) We have $P_0^*(s) = \dfrac{1}{s+\lambda_0}$. Solving recursively
$$P_1^*(s) = \dfrac{\lambda_0}{(s+\lambda_0)(s+\lambda_1)}$$

In general, we have that
$$P_k^*(s) = \dfrac{\prod_{i=0}^{k-1}\lambda_i}{\prod_{i=0}^{k}(s+\lambda_i)} \quad k \geq 0$$

(b) For the case $\lambda_i = \lambda$ $(i=0,1,2,\ldots)$ from part (a),
$$P_k^*(s) = \dfrac{\lambda^k}{(s+\lambda)^{k+1}} \quad k \geq 0$$

From Entry 10 Table I.4 we have that
$$\dfrac{1}{(s+\lambda)^{k+1}} \Leftrightarrow \dfrac{t^k}{k!}e^{-\lambda t}\delta(t)$$

and so
$$P_k^*(s) = \dfrac{\lambda^k}{(s+\lambda)^{k+1}} \Leftrightarrow \dfrac{(\lambda t)^k}{k!}e^{-\lambda t}\delta(t)$$

Hence, for $t \geq 0$,
$$P_k(t) = \dfrac{(\lambda t)^k}{k!}e^{-\lambda t} \qquad \blacksquare$$

PROBLEM 2.17.

Consider a time interval $(0, t)$ during which a Poisson process generates arrivals at an average rate λ. Derive Eq. (2.147) by considering the two events: exactly $k-1$ arrivals occur in the interval $(0, t-\Delta t)$ and the event that exactly one arrival occurs in the interval $(t-\Delta t, t)$. Considering the limit as $\Delta t \to 0$ we immediately arrive at our desired result.

2.16. – 2.17.

SOLUTION

Recall that the random variable of interest in Eq. (2.147) is X, the time interval required to collect k arrivals from a Poisson process, with $f_X(t)$ its pdf. To find $f_X(t)$, we first note that $f_X(t)\, \Delta t$ is the probability that k arrivals occur in $(0, t)$ with the kth arrival in $(t - \Delta t, t)$. Thus $f_X(t)\, \Delta t$ is simply the probability that there are $k - 1$ arrivals in $(0, t - \Delta t)$ and one arrival in $(t - \Delta t, t)$. Therefore we have

$$f_X(t)\, \Delta t = \left[\frac{[\lambda(t - \Delta t)]^{k-1}}{(k-1)!} e^{-\lambda(t - \Delta t)} \right] \left[\lambda \Delta t\, e^{-\lambda \Delta t} \right]$$

$$= \frac{\lambda[\lambda(t - \Delta t)]^{k-1}}{(k-1)!} e^{-\lambda t}\, \Delta t$$

Hence, as $\Delta t \to 0$, we obtain

$$f_X(t) = \frac{\lambda(\lambda t)^{k-1}}{(k-1)!} e^{-\lambda t}$$

PROBLEM 2.18.

A barber opens up for business at $t = 0$. Customers arrive at random in a Poisson fashion; that is, the pdf of interarrival time is $a(t) = \lambda e^{-\lambda t}$. Each haircut takes X sec (where X is some random variable). Find the probability P that the second arriving customer will not have to wait and also find W, the average value of his waiting time for the two following cases:
i. $X = c = $ constant.
ii. X is exponentially distributed with pdf:

$$b(x) = \mu e^{-\mu x}$$

SOLUTION

(i) $X = c = $ constant
The probability the second arriving customer will not have to wait is the probability that his interarrival time is $\geq c$.

$$P[\text{no waiting}] = \int_c^\infty \lambda e^{-\lambda t}\, dt = e^{-\lambda c} \qquad \blacksquare$$

The second customer will wait $c-y$ seconds if his interarrival time y is $\leq c$. Otherwise he does not wait. Hence his average waiting time is:

$$W = \int_0^c (c-y) P[y < \text{interarrival time} \leq y+dy] = \int_0^c (c-y)\lambda e^{-\lambda y}\,dy$$

$$W = c - \frac{1}{\lambda}[1-e^{-\lambda c}] \qquad \blacksquare$$

(ii) X is exponentially distributed with pdf $b(x) = \mu e^{-\mu x}$
As in (i) above, the second customer does not have to wait iff his interarrival time is $\geq$ the first customer's service time.

$P[\text{no waiting}] = P[\text{interarrival time} \geq \text{service time}]$

$$= \int_0^\infty P[\text{interarrival time} \geq \text{service time} \mid \text{service time} = x]$$

$$\cdot \mu e^{-\mu x}\,dx$$

$$= \int_0^\infty e^{-\lambda x} \mu e^{-\mu x}\,dx = \frac{\mu}{\lambda + \mu} \qquad \blacksquare$$

The average waiting time given that the service time was x seconds is $x - \frac{1}{\lambda}[1-e^{-\lambda x}]$ from case (i). Unconditioning on the service time gives

$$W = \int_0^\infty \left\{ x - \frac{1}{\lambda}[1-e^{-\lambda x}] \right\} \cdot \mu e^{-\mu x}\,dx$$

$$W = \frac{\lambda}{\mu(\mu+\lambda)} \qquad \blacksquare$$

PROBLEM 2.19.

At $t=0$ customer A places a request for service and finds all m servers busy and n other customers waiting for service in an M/M/m queueing system. All customers wait as long as necessary for service, waiting customers are served in order of arrival, and no new requests for service are permitted after $t=0$. Service times are assumed to be mutually independent, identical, exponentially distributed random variables, each with mean duration $1/\mu$.
(a) Find the expected length of time customer A spends waiting for service in the queue.
(b) Find the expected length of time from the arrival of customer A at $t=0$ until the system becomes completely empty (all customers complete service).

(c) Let X be the order of completion of service of customer A; that is, $X = k$ if A is the kth customer to complete service after $t = 0$. Find $P[X = k]$ $(k = 1, 2, \ldots, m+n+1)$.

(d) Find the probability that customer A completes service before the customer immediately ahead of him in the queue.

(e) Let $\tilde{w}$ be the amount of time customer A waits for service. Find $P[\tilde{w} > x]$.

SOLUTION

(a) When all m servers are busy, an interdeparture time has mean $1/m\mu$. Customer A must wait in the queue until $n+1$ such departures occur. Hence
$$E[\text{waiting time for A}] = \frac{n+1}{m\mu}$$ ∎

(b)
$$E[\text{time to empty system}] = \frac{n+1}{m\mu} + \frac{1}{m\mu} + \cdots + \frac{1}{\mu}$$
$$= \frac{n+1}{m\mu} + \frac{1}{\mu}\sum_{k=1}^{m}\frac{1}{k}$$ ∎

(c) Since $n+1$ customers must leave before A enters service, we have
$$P[X = k] = 0 \quad \text{for } k = 1, 2, \ldots, n, n+1$$ ∎

For $n+2 \leq k \leq m+n+1$, A's departure order is equally likely to be any of the m possible values (by symmetry and the memoryless property). Thus
$$P[X = k] = \frac{1}{m} \quad \text{for } k = n+2, \ldots, m+n+1$$ ∎

(d) Let B be the customer immediately ahead of A in the queue. A completes service before B if A enters service before B finishes (which occurs with probability $\frac{m-1}{m}$) and, once in service, A finishes before B (which occurs with probability $\frac{1}{2}$). Hence
$$P[\text{A completes service before B}] = \frac{m-1}{2m}$$ ∎

(e) $\tilde{w} = \tilde{d}_1 + \cdots + \tilde{d}_{n+1}$ where $\tilde{d}_i$ are interdeparture times, which are independent and identically distributed from an exponential distribution with mean $1/m\mu$.

2.19.

$\therefore \tilde{d}_1 + \cdots + \tilde{d}_{n+1}$ has an Erlang-$(n+1)$ distribution with parameter $m\mu$. That is, $\tilde{w}$ has pdf

$$f_{\tilde{w}}(x) = \frac{m\mu(m\mu x)^n}{n!} e^{-m\mu x}.$$

Integrating (by parts repeatedly) we get

$$P[\tilde{w} > x] = \sum_{i=0}^{n} \frac{(m\mu x)^i}{i!} e^{-m\mu x} \qquad \blacksquare$$

PROBLEM 2.20.†

In this problem we wish to proceed from Eq. (2.162) to the transient solution in Eq. (2.163). Since $P^*(z, s)$ must converge in the region $|z| \leq 1$ for $\text{Re}(s) > 0$, then, in this region, the zeros of the denominator in Eq. (2.162) must also be zeros of the numerator.

(a) Find those two values of z that give the denominator zeros, and denote them by $\alpha_1(s)$, $\alpha_2(s)$ where $|\alpha_2(s)| < |\alpha_1(s)|$.

(b) Using Rouche's theorem (see Appendix I) show that the denominator of $P^*(z, s)$ has a single zero within the unit disk $|z| \leq 1$.

(c) Requiring that the numerator of $P^*(z, s)$ vanish at $z = \alpha_2(s)$ from our earlier considerations, find an explicit expression for $P_0^*(s)$.

(d) Write $P^*(z, s)$ in terms of $\alpha_1(s) = \alpha_1$ and $\alpha_2(s) = \alpha_2$. Then show that this equation may be reduced to

$$P^*(z, s) = \frac{(z^i + \alpha_2 z^{i-1} + \cdots + \alpha_2^i) + \alpha_2^{i+1}/(1-\alpha_2)}{\lambda \alpha_1 (1 - z/\alpha_1)}$$

(e) For $k \geq i$, using the fact that $|\alpha_2| < 1$ and that $\alpha_1 \alpha_2 = \mu/\lambda$ show that the inversion on z yields the following expression for $P_k^*(s)$, which is the Laplace transform for our transient probabilities $P_k(t)$:

$$P_k^*(s) = \frac{1}{\lambda}\left[\alpha_1^{i-k-1} + \left(\frac{\mu}{\lambda}\right)\alpha_1^{i-k-3} + \left(\frac{\mu}{\lambda}\right)^2 \alpha_1^{i-k-5} + \cdots \right.$$

$$\left. + \left(\frac{\mu}{\lambda}\right)^i \alpha_1^{-i-k-1} + \left(\frac{\lambda}{\mu}\right)^{k+1} \sum_{j=k+i+2}^{\infty} \left(\frac{\mu}{\lambda \alpha_1}\right)^j \right]$$

(f) In what follows we take advantage of property 5 in Table I.3 and also we make use of the following transform pair:

2.19.–2.20.

$$k\rho^{k/2}t^{-1}I_k(at) \Leftrightarrow \left[\frac{s+\sqrt{s^2-4\lambda\mu}}{2\lambda}\right]^{-k}$$

where ρ and a are as defined in Eqs. (2.164), (2.165) and where $I_k(x)$ is the modified Bessel function of the first kind of order k as defined in Eq. (2.166). Using these facts and the simple relations among Bessel functions, namely,

$$\frac{2k}{x}I_k(x) = I_{k-1}(x) - I_{k+1}(x) \quad \text{and} \quad I_k(x) = I_{-k}(x)$$

show that Eq. (2.163) is the inverse transform for the expression shown in part (e), thus establishing the transient solution for $k \geq i$.

(g) Starting with the equation in part (d), extend the applicability of Eq. (2.163) to the range $k < i$.

SOLUTION

(a) Eq. (2.162) is

$$P^*(z,s) = \frac{z^{i+1} - \mu(1-z)P_0^*(s)}{sz - (1-z)(\mu-\lambda z)} = \frac{N(z,s)}{D(z,s)} = \frac{N}{D}$$

We may write $D = -[\lambda z^2 - (s+\mu+\lambda)z + \mu]$. The two zeros of D are

$$\alpha_1(s) = \frac{s+\mu+\lambda+\sqrt{(s+\mu+\lambda)^2-4\lambda\mu}}{2\lambda}$$

and

$$\alpha_2(s) = \frac{s+\mu+\lambda-\sqrt{(s+\mu+\lambda)^2-4\lambda\mu}}{2\lambda}$$

Thus $D = -\lambda[z-\alpha_1(s)][z-\alpha_2(s)]$. We now wish to show that $|\alpha_2(s)| < |\alpha_1(s)|$ for $\text{Re}(s) > 0$, or that $|\alpha_1(s)|^2 - |\alpha_2(s)|^2 > 0$. Defining $h \triangleq \sqrt{(s+\mu+\lambda)^2 - 4\lambda\mu}$ and substituting for $\alpha_1(s)$ and $\alpha_2(s)$ in the latter inequality yields the equivalent condition

(*) $\quad \text{Re}(s+\mu+\lambda)\text{Re}(h) + \text{Im}(s+\mu+\lambda)\text{Im}(h) > 0$

Thus we need only show (*) for $\text{Re}(s) > 0$. There are three cases to consider.

Case (1): $\text{Im}(s) = 0$
We may express h as

$$h = \sqrt{s^2 + 2s(\mu+\lambda) + (\mu+\lambda)^2 - 4\lambda\mu} = \sqrt{s^2 + 2s(\mu+\lambda) + (\mu-\lambda)^2}$$

However, in this case we note that s is a positive real number. Therefore h is also real and positive, and so (*) holds.

2.20.

Case (2): $\text{Im}(s) > 0$

In this case $\text{Re}(s+\mu+\lambda) > 0$ and $\text{Im}(s+\mu+\lambda) > 0$. Thus $s+\mu+\lambda$ may be represented as a point in the first quadrant of the complex plane. The complex number $(s+\mu+\lambda)^2$ must therefore be in the first or second quadrant. Subtracting a real number to form $(s+\mu+\lambda)^2 - 4\lambda\mu = h^2$ leaves us in the first or second quadrant, and taking the square root yields a point (h) in the first quadrant. That is, $\text{Re}(h) > 0$ and $\text{Im}(h) > 0$. Thus we see that (*) is satisfied.

Case (3): $\text{Im}(s) < 0$

In this case $\text{Re}(s+\mu+\lambda) > 0$ and $\text{Im}(s+\mu+\lambda) < 0$. In a manner similar to that used in Case (2) above, we find that $\text{Re}(h) > 0$ and $\text{Im}(h) < 0$. Thus once again (*) holds.

Since (*) is satisfied in all three cases, we have shown that, for $\text{Re}(s) > 0$, $|\alpha_2(s)| < |\alpha_1(s)|$ as desired.

(b) Decompose D as follows: $D = f_s(z) + g(z)$ where $f_s(z) = (s+\mu+\lambda)z$, $g(z) = -(\lambda z^2 + \mu)$. Then for $|z| = 1$ (and $\text{Re}(s) > 0$)

$$|f_s(z)| = |\lambda+\mu+s| \geq \lambda+\mu+\text{Re}(s) > \lambda+\mu$$

and

$$|g(z)| \leq \lambda+\mu.$$

Thus, on $|z| = 1$, $|g(z)| < |f_s(z)|$. Also, $f_s(z)$ and $g(z)$ are analytic inside and on $|z| = 1$. Thus by Rouche's theorem, as $f_s(z)$ has one zero in the range $|z| < 1$ (at $z = 0$), so does $D = f_s(z) + g(z)$.

(c) Since $|\alpha_2(s)| < |\alpha_1(s)|$, the one zero of D of interest must be $\alpha_2(s)$. By the analyticity of $P^*(z, s)$ for $|z| < 1$ ($\text{Re}(s) > 0$), $\alpha_2(s)$ is a zero of the numerator. Hence

$$[\alpha_2(s)]^{i+1} - \mu[1-\alpha_2(s)]P_0^*(s) = 0$$

$$\therefore P_0^*(s) = \frac{[\alpha_2(s)]^{i+1}}{\mu[1-\alpha_2(s)]} \qquad \blacksquare$$

(d)

$$P^*(z, s) = \frac{z^{i+1} - \mu(1-z)\dfrac{\alpha_2^{i+1}}{\mu(1-\alpha_2)}}{-\lambda(z-\alpha_1)(z-\alpha_2)}$$

$$= \frac{(1-\alpha_2)z^{i+1} - (1-z)\alpha_2^{i+1}}{-\lambda(z-\alpha_1)(z-\alpha_2)(1-\alpha_2)}$$

The numerator can be written as

2.20.

$$z^{i+1} - \alpha_2{}^{i+1} - \alpha_2 z(z^i - \alpha_2{}^i) = (z - \alpha_2)(z^i + z^{i-1}\alpha_2 + \cdots + z\alpha_2{}^{i-1} + \alpha_2{}^i)$$

$$- \alpha_2 z(z - \alpha_2)(z^{i-1} + z^{i-2}\alpha_2 + \cdots + z\alpha_2{}^{i-2} + \alpha_2{}^{i-1})$$

$$= (z - \alpha_2)[(1 - \alpha_2)(z^i + z^{i-1}\alpha_2 + \cdots + z\alpha_2{}^{i-1} + \alpha_2{}^i) + \alpha_2{}^{i+1}]$$

Thus

$$P^*(z,s) = \frac{(z - \alpha_2)[(1 - \alpha_2)(z^i + \alpha_2 z^{i-1} + \cdots + \alpha_2{}^i) + \alpha_2{}^{i+1}]}{-\lambda(z - \alpha_2)(z - \alpha_1)(1 - \alpha_2)}$$

$$P^*(z,s) = \frac{(z^i + \alpha_2 z^{i-1} + \cdots + \alpha_2{}^i) + \alpha_2{}^{i+1}/(1 - \alpha_2)}{\lambda \alpha_1 (1 - z/\alpha_1)}$$

(e) We may rewrite

$$P^*(z,s) = \frac{(z^i + \alpha_2 z^{i-1} + \cdots + \alpha_2{}^i) + \alpha_2{}^{i+1}/(1 - \alpha_2)}{\lambda \alpha_1} \sum_{k=0}^{\infty} \left(\frac{z}{\alpha_1}\right)^k$$

From this power series, we recognize for $k \geq i$,

$$P_k^*(s) = \frac{1}{\lambda \alpha_1}\left[\left(\frac{1}{\alpha_1}\right)^{k-i} + \alpha_2\left(\frac{1}{\alpha_1}\right)^{k-i+1} + \cdots + \alpha_2{}^i\left(\frac{1}{\alpha_1}\right)^k\right]$$

$$+ \frac{\alpha_2{}^{i+1}}{1 - \alpha_2} \cdot \frac{1}{\lambda \alpha_1{}^{k+1}}$$

Now, using the fact that $|\alpha_2| < 1$,

$$P_k^*(s) = \frac{1}{\lambda \alpha_1}\left[\left(\frac{1}{\alpha_1}\right)^{k-i} + \alpha_2\left(\frac{1}{\alpha_1}\right)^{k-i+1} + \cdots + \alpha_2{}^i\left(\frac{1}{\alpha_1}\right)^k\right]$$

$$+ \frac{1}{\lambda} \frac{\alpha_2{}^{i+1}}{\alpha_1{}^{k+1}}\left[1 + \alpha_2 + \alpha_2{}^2 + \cdots\right]$$

and thus

$$P_k^*(s) = \frac{1}{\lambda}\left[\left(\frac{1}{\alpha_1}\right)^{k-i+1} + \alpha_1\alpha_2\left(\frac{1}{\alpha_1}\right)^{k-i+3} + \cdots + (\alpha_1\alpha_2)^i\left(\frac{1}{\alpha_1}\right)^{k+i+1}\right]$$

$$+ \frac{1}{\lambda} \frac{\alpha_2{}^{k+i+2}}{(\alpha_1\alpha_2)^{k+1}}\left[1 + \alpha_2 + \alpha_2{}^2 + \cdots\right]$$

Since $\alpha_1\alpha_2 = \frac{\mu}{\lambda}$, we have $\alpha_2 = \frac{\mu}{\lambda\alpha_1}$. Thus for $k \geq i$,

2.20.

$$P_k^*(s) = \frac{1}{\lambda}\left[\alpha_1^{i-k-1} + \left(\frac{\mu}{\lambda}\right)\alpha_1^{i-k-3} + \left(\frac{\mu}{\lambda}\right)^2\alpha_1^{i-k-5} + \cdots\right.$$

$$\left. + \left(\frac{\mu}{\lambda}\right)^i \alpha_1^{-i-k-1} + \left(\frac{\lambda}{\mu}\right)^{k+1} \sum_{j=k+i+2}^{\infty}\left(\frac{\mu}{\lambda\alpha_1}\right)^j\right]$$

(f) From property 5 in Table I.3 and the transform pair given in part (f) of the exercise statement, we have

$$\alpha_1^{-k} = \left[\frac{s+\mu+\lambda+\sqrt{(s+\mu+\lambda)^2-4\lambda\mu}}{2\lambda}\right]^{-k} \Leftrightarrow e^{-(\lambda+\mu)t}k\rho^{k/2}t^{-1}I_k(at)$$

where $\rho = \lambda/\mu$ and $a = 2\mu\rho^{1/2} = 2\sqrt{\lambda\mu}$. Inverting the expression for $P_k^*(s)$ from part (e), using the fact that

$$\frac{2k}{at}I_k(at) = I_{k-1}(at) - I_{k+1}(at),$$

and carefully reducing the result, will yield Eq. (2.163) for $k \geq i$. Details can be found in the book *Elements of Queueing Theory* by T. L. Saaty, pp 91-93 (McGraw-Hill 1961).

(g) For $k < i$, the value of $P_k^*(s)$ must be found, and then it must be inverted. Further details are on page 93 in the above-mentioned book by Saaty.

PROBLEM 2.21.

The random variables $X_1, X_2, \ldots, X_i, \ldots$ are independent, identically distributed random variables each with density $f_X(x)$ and characteristic function $\phi_X(u) = E[e^{juX}]$. Consider a Poisson process $N(t)$ with parameter λ which is independent of the random variables X_i. Consider now a second random process of the form

$$X(t) = \sum_{i=1}^{N(t)} X_i$$

This second random process is clearly a family of staircase functions where the jumps occur at the discontinuities of the random process $N(t)$; the magnitudes of such jumps are given by the random variables X_i. Show that the characteristic function of this second random process is given by

$$\phi_{X(t)}(u) = e^{\lambda t[\phi_X(u)-1]}$$

SOLUTION

Condition on $N(t) = n$. Then the (conditional) characteristic function is given as follows:

$$\phi_{X(t)}(u \mid N(t) = n) = E[e^{juX(t)} \mid N(t) = n] = E\left[e^{ju\sum_{i=1}^{n} X_i}\right]$$

$$= \left(E\left[e^{juX_1}\right]\right)^n \quad \text{(as the } X_i \text{ are i.i.d.)}$$

$$= \left(\phi_X(u)\right)^n$$

Since $N(t)$ is Poisson distributed with parameter λt, we may uncondition as follows (note that we assume $P[N(0) = 0] = 1$):

$$\phi_{X(t)}(u) = \sum_{n=0}^{\infty} \phi_{X(t)}(u \mid N(t) = n) \cdot P[N(t) = n]$$

$$= \sum_{n=0}^{\infty} \left(\phi_X(u)\right)^n \frac{(\lambda t)^n e^{-\lambda t}}{n!} = e^{-\lambda t} e^{\lambda t \phi_X(u)}$$

$$= e^{\lambda t [\phi_X(u) - 1]}$$

PROBLEM 2.22.

Passengers and taxis arrive at a service point from independent Poisson processes at rates λ, μ, respectively. Let the queue size at time t be q_t, a negative value denoting a line of taxis, a positive value denoting a queue of passengers. Show that, starting with $q_0 = 0$, the distribution of q_t is given by the difference between independent Poisson variables of means λt, μt. Show by using the normal approximation that if $\lambda = \mu$, the probability that $-k \leq q_t \leq k$ is, for large t, $(2k+1)(4\pi\lambda t)^{-1/2}$.

SOLUTION

Let $P_n(t) = P[q_t = n]$, $-\infty < n < \infty$. For $n < 0$, we have n taxis waiting. For $n > 0$, we have n passengers waiting. For $n = 0$, we have no queue. The equations of motion are simply

$$\frac{d P_n(t)}{dt} = -(\lambda + \mu) P_n(t) + \mu P_{n+1}(t) + \lambda P_{n-1}(t) \quad \text{for } -\infty < n < \infty$$

Define $P(z, t) = \sum_{-\infty}^{+\infty} P_n(t) z^n$. Multiplying the nth equation by z^n and summing yields

$$\sum_{-\infty}^{+\infty} \frac{d P_n(t)}{dt} z^n = -(\lambda + \mu) \sum_{-\infty}^{+\infty} P_n(t) z^n + \mu \sum_{-\infty}^{+\infty} P_{n+1}(t) z^n + \lambda \sum_{-\infty}^{+\infty} P_{n-1}(t) z^n$$

$$\frac{\partial P(z, t)}{\partial t} = -P(z, t) \left[\lambda(1 - z) + \mu\left(1 - \frac{1}{z}\right) \right]$$

Solving this partial differential equation gives

$$P(z, t) = A(z) e^{-\lambda t(1 - z) - \mu t\left(1 - \frac{1}{z}\right)}$$

The initial condition $P_0(0) = 1$ implies $P(z, 0) = 1$. Hence $A(z) = 1$.

$$\therefore \quad P(z, t) = e^{-\lambda t(1 - z) - \mu t\left(1 - \frac{1}{z}\right)}$$

But the arrival process for passengers has, for fixed t, the z-transform

$$P_1(z, t) = \sum_{n=0}^{\infty} \frac{(\lambda t)^n}{n!} e^{-\lambda t} z^n = e^{-\lambda t(1 - z)}$$

and the arrival process for taxis has the corresponding transform

$$P_2(z, t) = \sum_{n=0}^{\infty} \frac{(\mu t)^n}{n!} e^{-\mu t} z^n = e^{-\mu t(1 - z)}$$

From these results we see that

$$P(z, t) = P_1(z, t) \, P_2\!\left(\frac{1}{z}, t\right)$$

Now note for independent random variables X, Y (whose z-transforms are F and G respectively) and with $W = X - Y$, that

$$H(z) = E[z^W] = E[z^{X-Y}] = E[z^X] \cdot E[z^{-Y}]$$

$$= E[z^X] \cdot E\!\left[\left(\frac{1}{z}\right)^Y\right] = F(z) \, G\!\left(\frac{1}{z}\right)$$

2.22.

Thus $P_1(z, t) \cdot P_2\left(\frac{1}{z}, t\right)$ represents the z-transform of the difference between independent Poisson variables of means λt, μt. But since it is also equal to the z-transform $P(z, t)$ of q_t, we see that the distribution of q_t has the required form. We recognize that the mean of q_t is $\lambda t - \mu t$, and the variance is $\lambda t + \mu t$. Now assume $\lambda = \mu$. Hence the mean of q_t is 0 with variance $2\lambda t$. Using the normal approximation, we have

$$P[-k \leqslant q_t \leqslant k] \cong \int_{-(k+\frac{1}{2})}^{k+\frac{1}{2}} \frac{1}{\sqrt{2\pi}\sqrt{2\lambda t}} e^{-\frac{x^2}{4\lambda t}} dx$$

But for large t, $e^{-\frac{x^2}{4\lambda t}} \to 1$. So

$$P[-k \leqslant q_t \leqslant k] \cong \int_{-(k+\frac{1}{2})}^{k+\frac{1}{2}} \frac{1}{\sqrt{4\pi \lambda t}} dx = \frac{2k+1}{\sqrt{4\pi \lambda t}}$$

2.22.

Chapter 3

Birth—Death Queueing Systems in Equilibrium

PROBLEM 3.1.

Consider a pure Markovian queueing system in which

$$\lambda_k = \begin{cases} \lambda & 0 \leq k \leq K \\ 2\lambda & K < k \end{cases}$$

$$\mu_k = \mu \qquad k = 1, 2, \ldots$$

(a) Find the equilibrium probabilities p_k for the number in the system.
(b) What relationship must exist among the parameters of the problem in order that the system be stable and, therefore, that this equilibrium solution in fact be reached? Interpret this answer in terms of the possible dynamics of the system.

SOLUTION

(a) Case (1): $0 \leq k \leq K+1$
Eq (3.11) gives

$$p_k = p_0 \prod_{i=0}^{k-1} \frac{\lambda}{\mu} = p_0 \left(\frac{\lambda}{\mu}\right)^k$$

Case (2): $k > K+1$
Eq (3.11) gives

$$p_k = p_0 \prod_{i=0}^{K} \frac{\lambda}{\mu} \prod_{i=K+1}^{k-1} \frac{2\lambda}{\mu}$$

$$p_k = p_0 \left[\left(\frac{\lambda}{\mu}\right)^{K+1} \left(\frac{2\lambda}{\mu}\right)^{k-K-1}\right] = p_0 \frac{1}{2^{K+1}} \left(\frac{2\lambda}{\mu}\right)^k$$

Using the conservation relation in Eq. (3.5) we solve for p_0 as follows:

3.1.

$$1 = p_0 \left[\sum_{k=0}^{K+1} \left(\frac{\lambda}{\mu}\right)^k + \sum_{k=K+2}^{\infty} \frac{1}{2^{K+1}} \left(\frac{2\lambda}{\mu}\right)^k \right]$$

$$= p_0 \left[\frac{1 - \left(\frac{\lambda}{\mu}\right)^{K+2}}{1 - \frac{\lambda}{\mu}} + \frac{1}{2^{K+1}} \left(\frac{2\lambda}{\mu}\right)^{K+2} \frac{1}{1 - \frac{2\lambda}{\mu}} \right]$$

Thus

$$p_0 = \frac{\left(1 - \frac{\lambda}{\mu}\right)\left(1 - \frac{2\lambda}{\mu}\right)}{1 - \frac{2\lambda}{\mu} + \left(\frac{\lambda}{\mu}\right)^{K+2}} \qquad \blacksquare$$

and

$$p_k = \begin{cases} \dfrac{\left(1 - \frac{\lambda}{\mu}\right)\left(1 - \frac{2\lambda}{\mu}\right)}{1 - \frac{2\lambda}{\mu} + \left(\frac{\lambda}{\mu}\right)^{K+2}} \left(\frac{\lambda}{\mu}\right)^k & 0 \leq k \leq K+1 \\[2em] \dfrac{\left(1 - \frac{\lambda}{\mu}\right)\left(1 - \frac{2\lambda}{\mu}\right)}{1 - \frac{2\lambda}{\mu} + \left(\frac{\lambda}{\mu}\right)^{K+2}} \left(\frac{2\lambda}{\mu}\right)^k \frac{1}{2^{K+1}} & k > K+1 \end{cases} \qquad \blacksquare$$

(b) We must have $\frac{2\lambda}{\mu} < 1$ or $2\lambda < \mu$ to insure that $S_1 < \infty$ and $S_2 = \infty$. We observe that if the system goes unstable, then $k > K$ with probability one; thus the relevant birth parameter is $\lambda_k = 2\lambda$ and our stability condition is $2\lambda < \mu$.

PROBLEM 3.2.

Consider a Markovian queueing system in which
$$\lambda_k = \alpha^k \lambda \qquad k \geq 0, \ 0 \leq \alpha < 1$$
$$\mu_k = \mu \qquad k \geq 1$$

(a) Find the equilibrium probability p_k of having k customers in the system. Express your answer in terms of p_0.

(b) Give an expression for p_0.

3.1. – 3.2.

SOLUTION

(a) From Eq. (3.11) we have

$$p_k = p_0 \prod_{i=0}^{k-1} \alpha^i \left(\frac{\lambda}{\mu}\right) = p_0 \left(\frac{\lambda}{\mu}\right)^k \alpha^{\sum_{i=0}^{k-1} i}$$

$$p_k = p_0 \left(\frac{\lambda}{\mu}\right)^k \alpha^{\frac{(k-1)k}{2}}$$

$$\therefore \; p_k = p_0 \left[\frac{\lambda \alpha^{(k-1)/2}}{\mu}\right]^k \quad \blacksquare$$

(b)

$$\sum_{k=0}^{\infty} p_k = 1 = p_0 \sum_{k=0}^{\infty} \left[\frac{\lambda \alpha^{(k-1)/2}}{\mu}\right]^k$$

So

$$p_0 = \frac{1}{\sum_{k=0}^{\infty} \left[\frac{\lambda \alpha^{(k-1)/2}}{\mu}\right]^k} \quad \blacksquare$$

Note for $0 \leq \alpha < 1$, this system is *always* stable.

PROBLEM 3.3.

Consider an M/M/2 queueing system where the average arrival rate is λ customers per second and the average service time is $1/\mu$ sec, where $\lambda < 2\mu$.

(a) Find the differential equations that govern the time-dependent probabilities $P_k(t)$.

(b) Find the equilibrium probabilities

$$p_k = \lim_{t \to \infty} P_k(t)$$

SOLUTION

$$\lambda_k = \lambda \quad k \geq 0, \qquad \mu_k = \begin{cases} \mu & k = 1 \\ 2\mu & k \geq 2 \end{cases}$$

(a)

$$\frac{dP_0(t)}{dt} = -\lambda P_0(t) + \mu P_1(t) \quad k = 0$$

$$\frac{dP_1(t)}{dt} = -(\lambda+\mu)P_1(t) + \lambda P_1(t) + 2\mu P_2(t) \quad k = 1$$

$$\frac{dP_k(t)}{dt} = -(\lambda+2\mu)P_k(t) + \lambda P_{k-1}(t) + 2\mu P_{k+1}(t) \quad k \geq 2$$

(b) Using Eq. (3.11) we have

$$p_k = p_0 \left(\frac{\lambda}{\mu}\right)\left(\frac{\lambda}{2\mu}\right)^{k-1} = 2 p_0 \left(\frac{\lambda}{2\mu}\right)^k \quad \text{for } k \geq 1$$

Using $\sum_{k=0}^{\infty} p_k = 1$ we have

$$p_0 + 2 p_0 \sum_{k=1}^{\infty} \left(\frac{\lambda}{2\mu}\right)^k = 1$$

$$p_0 \left[1 + \frac{\lambda}{\mu} \frac{1}{1 - \frac{\lambda}{2\mu}}\right] = 1$$

$$p_0 = \frac{1 - \frac{\lambda}{2\mu}}{1 + \frac{\lambda}{2\mu}}$$

$$p_k = \frac{1 - \frac{\lambda}{2\mu}}{1 + \frac{\lambda}{2\mu}} \cdot 2\left(\frac{\lambda}{2\mu}\right)^k \quad k \geq 1$$

3.3.

PROBLEM 3.4.

Consider an M/M/1 system with parameters λ, μ in which customers are impatient. Specifically, upon arrival, customers estimate their queueing time w and then join the queue with probability $e^{-\alpha w}$ (or leave with probability $1 - e^{-\alpha w}$). The estimate is $w = k/\mu$ when the new arrival finds k in the system. Assume $0 \leqslant \alpha$.

(a) In terms of p_0, find the equilibrium probabilities p_k of finding k in the system. Give an expression for p_0 in terms of the system parameters.

(b) For $0 < \alpha$, $0 < \mu$ under what conditions will the equilibrium solution hold?

(c) For $\alpha \to \infty$, find p_k explicitly and find the average number in the system.

SOLUTION

$$\lambda_k = \lambda e^{-\frac{\alpha k}{\mu}}, \quad \mu_k = \mu$$

(a) Eq (3.11) gives

$$p_k = p_0 \prod_{i=0}^{k-1} \frac{\lambda e^{-\frac{\alpha i}{\mu}}}{\mu} = p_0 \left(\frac{\lambda}{\mu}\right)^k e^{-\frac{\alpha}{\mu}\sum_{i=0}^{k-1} i}$$

$$p_k = p_0 \left(\frac{\lambda}{\mu}\right)^k e^{-\frac{\alpha k(k-1)}{2\mu}} \quad \blacksquare$$

$$\sum_{k=0}^{\infty} p_k = 1 = p_0 \sum_{k=0}^{\infty} \left(\frac{\lambda}{\mu}\right)^k e^{-\frac{\alpha k(k-1)}{2\mu}}$$

$$p_0 = \frac{1}{\sum_{k=0}^{\infty} \left(\frac{\lambda}{\mu}\right)^k e^{-\frac{\alpha k(k-1)}{2\mu}}} \quad \blacksquare$$

(b)

$$S_1 = \sum_{k=1}^{\infty} \prod_{i=0}^{k-1} \frac{\lambda_i}{\mu_{i+1}} = \sum_{k=1}^{\infty} \left(\frac{\lambda}{\mu}\right)^k e^{-\frac{\alpha k(k-1)}{2\mu}}$$

For $0 < \alpha$, $0 < \mu$ this series converges and $S_1 < \infty$. Also we see that $S_2 = \infty$ and so the equilibrium solution holds *for all* $0 < \alpha$, $0 < \mu$.

(c) For $\alpha \to \infty$, $p_k \to 0$ for $k \geqslant 2$. Thus we move between only two states E_0 and E_1 with $\lambda_0 = \lambda$, $\lambda_1 = 0$ and $\mu_1 = \mu$. Thus, solving for p_0 and p_1 (see also Exercise 2.11 for $P_k(t)$) gives

3.4.

$$p_0 = \frac{\mu}{\lambda+\mu}$$ ∎

$$p_1 = \frac{\lambda}{\lambda+\mu}$$ ∎

$$\bar{N} = 0\cdot p_0 + 1\cdot p_1 = \frac{\lambda}{\lambda+\mu}$$ ∎

PROBLEM 3.5.

Consider a birth-death system with the following birth and death coefficients:
$$\lambda_k = (k+2)\lambda \quad k = 0,1,2,\ldots$$
$$\mu_k = k\mu \quad k = 1,2,3,\ldots$$
All other coefficients are zero.
(a) Solve for p_k. Be sure to express your answer explicitly in terms of λ, k, and μ only.
(b) Find the average number of customers in the system.

SOLUTION

(a)

$$p_k = p_0 \left(\frac{\lambda}{\mu}\right)^k \frac{2\cdot 3 \cdots (k+1)}{1\cdot 2 \cdots k}$$

$$p_k = p_0(k+1)\left(\frac{\lambda}{\mu}\right)^k \quad \text{for } k \geq 0$$

$$1 = \sum_{k=0}^{\infty} p_k = p_0 \sum_{k=0}^{\infty}(k+1)\left(\frac{\lambda}{\mu}\right)^k$$

Here we demonstrate the "differentiation trick" for summing series (similar to that on page 69). Since

$$\sum_{k=0}^{\infty}(k+1)\left(\frac{\lambda}{\mu}\right)^k = \frac{\partial}{\partial\left(\frac{\lambda}{\mu}\right)} \sum_{k=0}^{\infty}\left(\frac{\lambda}{\mu}\right)^{k+1}$$

we have

$$\sum_{k=0}^{\infty}(k+1)\left(\frac{\lambda}{\mu}\right)^k = \frac{\partial}{\partial\left(\frac{\lambda}{\mu}\right)}\left[\frac{\frac{\lambda}{\mu}}{1-\frac{\lambda}{\mu}}\right] = \frac{1}{\left(1-\frac{\lambda}{\mu}\right)^2}$$

Thus

$$p_0 = \left(1-\frac{\lambda}{\mu}\right)^2 \qquad ∎$$

and so

$$p_k = \left(1-\frac{\lambda}{\mu}\right)^2 (k+1)\left(\frac{\lambda}{\mu}\right)^k \quad k \geq 0 \qquad ∎$$

(b)

$$\overline{N} = \sum_{k=0}^{\infty} k p_k = \sum_{k=1}^{\infty} k\left(1-\frac{\lambda}{\mu}\right)^2 (k+1)\left(\frac{\lambda}{\mu}\right)^k$$

$$= \left(1-\frac{\lambda}{\mu}\right)^2 \frac{\lambda}{\mu} \frac{\partial^2}{\partial\left(\frac{\lambda}{\mu}\right)^2}\left[\sum_{k=0}^{\infty}\left(\frac{\lambda}{\mu}\right)^{k+1}\right]$$

$$\overline{N} = \left(1-\frac{\lambda}{\mu}\right)^2 \frac{\lambda}{\mu} \frac{\partial}{\partial\left(\frac{\lambda}{\mu}\right)}\left[\frac{1}{\left(1-\frac{\lambda}{\mu}\right)^2}\right] = \left(1-\frac{\lambda}{\mu}\right)^2 \frac{\lambda}{\mu} \frac{2}{\left(1-\frac{\lambda}{\mu}\right)^3}$$

$$\overline{N} = \frac{2\left(\frac{\lambda}{\mu}\right)}{1-\frac{\lambda}{\mu}} \qquad ∎$$

PROBLEM 3.6.

Consider a birth-death process with the following coefficients:

$$\lambda_k = \alpha k(K_2 - k) \quad k = K_1, K_1+1, \ldots, K_2$$

$$\mu_k = \beta k(k - K_1) \quad k = K_1, K_1+1, \ldots, K_2$$

where $K_1 \leq K_2$ and where these coefficients are zero outside the range $K_1 \leq k \leq K_2$. Solve for p_k (assuming that the system initially contains $K_1 \leq k \leq K_2$ customers).

3.5.–3.6.

SOLUTION

Clearly $p_k = 0$ for $k < K_1$, $k > K_2$. Using the obvious translation of Eq. (3.11) in the range $K_1 \leq k \leq K_2$, we get

$$p_k = p_{K_1} \prod_{i=K_1}^{k-1} \frac{\alpha i (K_2 - i)}{\beta(i+1)(i+1-K_1)}$$

$$p_k = p_{K_1} \left(\frac{\alpha}{\beta}\right)^{k-K_1} \frac{K_1(K_1+1)\cdots(k-1)}{(K_1+1)(K_1+2)\cdots k} \cdot \frac{(K_2-K_1)\cdots(K_2-k+1)}{1\cdot 2\cdots(k-K_1)}$$

Multiplying the top and bottom of the right-hand expression by $K_1!(K_2-k)!$ we find

$$p_k = p_{K_1} \left(\frac{\alpha}{\beta}\right)^{k-K_1} \frac{(k-1)!\, K_1 (K_2-K_1)!}{(k-1)!\, k (K_2-k)!\, (k-K_1)!}$$

$$p_k = p_{K_1} \left(\frac{\alpha}{\beta}\right)^{k-K_1} \frac{K_1}{k} \binom{K_2-K_1}{k-K_1} \quad k = K_1, \ldots, K_2 \qquad \blacksquare$$

where, by conserving probability, we get

$$p_{K_1} = \frac{1}{\sum_{k=K_1}^{K_2} \left(\frac{\alpha}{\beta}\right)^{k-K_1} \frac{K_1}{k} \binom{K_2-K_1}{k-K_1}} \qquad \blacksquare$$

PROBLEM 3.7.

Consider an M/M/m system that is to serve the pooled sum of two Poisson arrival streams; the ith stream has an average arrival rate given by λ_i and exponentially distributed service times with mean $1/\mu_i$ ($i = 1, 2$). The first stream is an ordinary stream whereby each arrival requires exactly one of the m servers; if all m servers are busy then any newly arriving customer of type 1 is lost. Customers from the second class each require the simultaneous use of m_0 servers (and will occupy them all simultaneously for the same exponentially distributed amount of time whose mean is $1/\mu_2$ sec); if a customer from this class finds less than m_0 idle servers then he too is lost to the system. Find the fraction of type 1 customers and the fraction of type 2 customers that are lost.

SOLUTION

The state space consists of (a finite number of) two-dimensional states. $E_{i,j}$ represents the system state in which i customers of type 1 and j customers of type 2 are present. Let n and k be such that $m = nm_0 + k$, $0 \leq k \leq m_0 - 1$. The set of states consists of all $E_{i,j}$ such that $0 \leq i \leq m$, $0 \leq j \leq n$ and $i + jm_0 \leq m$ (each customer of type 2 requires m_0 servers). The state diagram is drawn below:

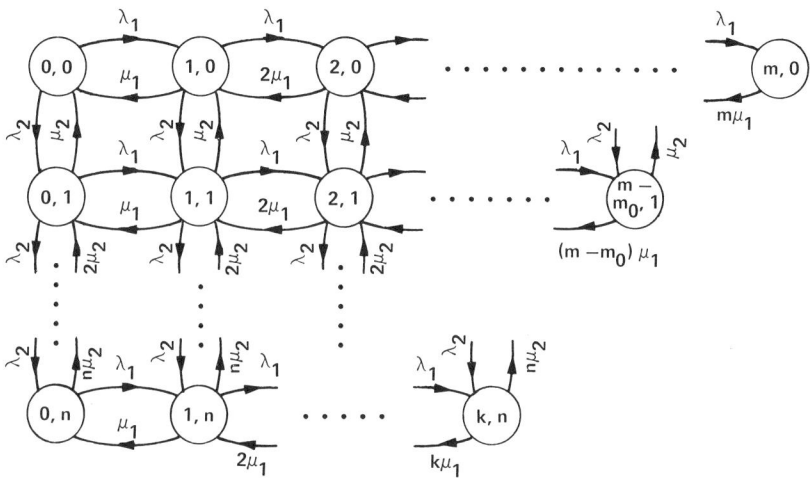

The general balance equation (Rate In = Rate Out) for an "internal" state is

$$\lambda_2 p(i,j-1) + \lambda_1 p(i-1,j) + (j+1)\mu_2 p(i,j+1) + (i+1)\mu_1 p(i+1,j)$$
$$= j\mu_2 p(i,j) + i\mu_1 p(i,j) + \lambda_2 p(i,j) + \lambda_1 p(i,j)$$

This set of equations, supplemented with the boundary equations and the conservation of probability, has a unique solution (since we have an ergodic Markov chain). Thus, if we can guess the $p(i,j)$ and show that they satisfy those equations, then we know we have the equilibrium solution. We may observe that each row and each column of our state diagram is the same as that in Fig. 3.9. Therefore, we choose to guess the two-dimensional "product" form solution as suggested by Eq. (3.45):

3.7.

$$p(i,j) = \frac{1}{i!}\left(\frac{\lambda_1}{\mu_1}\right)^i \frac{1}{j!}\left(\frac{\lambda_2}{\mu_2}\right)^j p(0,0)$$

(The local balance approach described in Section 4.8 leads us directly to this solution.) Checking this proposed solution for an internal state, we have

$$\frac{\text{Rate In}}{p(0,0)} = \lambda_2 \frac{1}{i!}\left(\frac{\lambda_1}{\mu_1}\right)^i \frac{1}{(j-1)!}\left(\frac{\lambda_2}{\mu_2}\right)^{j-1} + \lambda_1 \frac{1}{(i-1)!}\left(\frac{\lambda_1}{\mu_1}\right)^{i-1} \frac{1}{j!}\left(\frac{\lambda_2}{\mu_2}\right)^j$$

$$+ (j+1)\mu_2 \frac{1}{i!}\left(\frac{\lambda_1}{\mu_1}\right)^i \frac{1}{(j+1)!}\left(\frac{\lambda_2}{\mu_2}\right)^{j+1} + (i+1)\mu_1 \frac{1}{(i+1)!}\left(\frac{\lambda_1}{\mu_1}\right)^{i+1} \frac{1}{j!}\left(\frac{\lambda_2}{\mu_2}\right)^j$$

$$= (j\mu_2 + i\mu_1 + \lambda_2 + \lambda_1) \frac{1}{i!}\left(\frac{\lambda_1}{\mu_1}\right)^i \frac{1}{j!}\left(\frac{\lambda_2}{\mu_2}\right)^j$$

$$= \frac{\text{Rate Out}}{p(0,0)}$$

which confirms our proposed solution. (Note: the boundary states should also be checked.) Using conservation of probability to evaluate $p(0,0)$ we have the complete solution:

$$p(i,j) = \frac{\frac{1}{i!}\left(\frac{\lambda_1}{\mu_1}\right)^i \frac{1}{j!}\left(\frac{\lambda_2}{\mu_2}\right)^j}{\sum_{j=0}^{n} \frac{1}{j!}\left(\frac{\lambda_2}{\mu_2}\right)^j \sum_{i=0}^{m-jm_0} \frac{1}{i!}\left(\frac{\lambda_1}{\mu_1}\right)^i} \qquad \blacksquare$$

Let us now find the loss probabilities. A customer of type 1 is lost iff all m servers are busy. This occurs if the system is in the rightmost state of any row. Summing the probabilities of these $n+1$ states, we have

$$P[\text{type 1 lost}] = p(0,0)\left[\frac{1}{m!}\left(\frac{\lambda_1}{\mu_1}\right)^m + \frac{1}{(m-m_0)!}\left(\frac{\lambda_1}{\mu_1}\right)^{m-m_0}\left(\frac{\lambda_2}{\mu_2}\right) + \cdots\right.$$

$$\left. + \frac{1}{(m-nm_0)!}\left(\frac{\lambda_1}{\mu_1}\right)^{m-nm_0} \frac{1}{n!}\left(\frac{\lambda_2}{\mu_2}\right)^n\right]$$

$$P[\text{type 1 lost}] = \frac{\sum_{j=0}^{n} \frac{1}{j!}\left(\frac{\lambda_2}{\mu_2}\right)^j \frac{1}{(m-jm_0)!}\left(\frac{\lambda_1}{\mu_1}\right)^{m-jm_0}}{\sum_{j=0}^{n} \frac{1}{j!}\left(\frac{\lambda_2}{\mu_2}\right)^j \sum_{i=0}^{m-jm_0} \frac{1}{i!}\left(\frac{\lambda_1}{\mu_1}\right)^i} \qquad \blacksquare$$

3.7.

A type 2 customer is lost iff less than m_0 servers are idle, i.e. at least $m - m_0 + 1$ servers are busy.

$$P[\text{type 2 lost}] = \sum_{i=m-m_0+1}^{m} p(i,0) + \sum_{i=m-2m_0+1}^{m-m_0} p(i,1) + \cdots$$

$$+ \sum_{i=m-nm_0+1}^{m-(n-1)m_0} p(i,n-1) + \sum_{i=0}^{k} p(i,n)$$

Summing these $m+1$ probabilities we get

$$P[\text{type 2 lost}] = \frac{\sum_{j=0}^{n-1} \frac{1}{j!}\left(\frac{\lambda_2}{\mu_2}\right)^j \sum_{i=m-(j+1)m_0+1}^{m-jm_0} \frac{1}{i!}\left(\frac{\lambda_1}{\mu_1}\right)^i}{\sum_{j=0}^{n} \frac{1}{j!}\left(\frac{\lambda_2}{\mu_2}\right)^j \sum_{i=0}^{m-jm_0} \frac{1}{i!}\left(\frac{\lambda_1}{\mu_1}\right)^i}$$

$$+ \frac{\frac{1}{n!}\left(\frac{\lambda_2}{\mu_2}\right)^n \sum_{i=0}^{k} \frac{1}{i!}\left(\frac{\lambda_1}{\mu_1}\right)^i}{\sum_{j=0}^{n} \frac{1}{j!}\left(\frac{\lambda_2}{\mu_2}\right)^j \sum_{i=0}^{m-jm_0} \frac{1}{i!}\left(\frac{\lambda_1}{\mu_1}\right)^i} \quad \blacksquare$$

PROBLEM 3.8.

Consider a finite customer population system with a single server such as that considered in Section 3.8; let the parameters M, λ be replaced by M, λ'. It can be shown that if $M \to \infty$ and $\lambda' \to 0$ such that $\lim M\lambda' = \lambda$ then the finite population system becomes an infinite population system with exponential interarrival times (at a mean rate of λ customers per second). Now consider the case of Section 3.10; the parameters of that case are now to be denoted M, λ', m, μ, K in the obvious way. Show what value these parameters must take on if they are to represent the earlier cases described in Sections 3.2, 3.4, 3.5, 3.6, 3.7, 3.8, or 3.9.

SOLUTION

The solution to our M/M/m/K/M system is given in Eq. (3.51) and Eq. (3.52). We must specialize the parameters and take the limit $M\lambda' = \lambda$ as $M \to \infty$, $\lambda' \to 0$ of these equations for each case below:

(i) M/M/1 (section 3.2)

$$m = 1, \quad K = \infty, \quad M = \infty, \quad \lambda' = \lambda/M$$

Eq. (3.52) gives

$$p_k = p_0 \left(\frac{\lambda}{M\mu}\right)^k \binom{M}{k} k! = p_0 \left(\frac{\lambda}{\mu}\right)^k \frac{M \cdot (M-1) \cdots (M-k+1)}{M^k}$$

$$\lim_{M \to \infty} p_k = p_0 \left(\frac{\lambda}{\mu}\right)^k \quad \text{(as in Eq. (3.21))}$$

(ii) M/M/∞ (section 3.4)

$$m = \infty, \quad K = \infty, \quad M = \infty, \quad \lambda' = \lambda/M$$

Eq. (3.51) gives

$$p_k = p_0 \left(\frac{\lambda}{M\mu}\right)^k \binom{M}{k} = p_0 \left(\frac{\lambda}{\mu}\right)^k \frac{M \cdot (M-1) \cdots (M-k+1)}{M^k k!}$$

$$\lim_{M \to \infty} p_k = p_0 \left(\frac{\lambda}{\mu}\right)^k \frac{1}{k!} \quad \text{(as in Eq. (3.34))}$$

(iii) M/M/m (section 3.5)

$$m = m, \quad K = \infty, \quad M = \infty, \quad \lambda' = \lambda/M$$

For $k \leq m-1$, Eq. (3.51) gives

$$p_k = p_0 \left(\frac{\lambda}{M\mu}\right)^k \binom{M}{k} = p_0 \left(\frac{\lambda}{\mu}\right)^k \frac{M \cdot (M-1) \cdots (M-k+1)}{M^k k!}$$

$$\lim_{M \to \infty} p_k = p_0 \left(\frac{\lambda}{\mu}\right)^k \frac{1}{k!} = p_0 \frac{(m\rho)^k}{k!} \quad \text{where } \rho = \frac{\lambda}{m\mu} \quad \text{(as in Eq. (3.37))}$$

For $k \geq m$, Eq. (3.52) gives

$$p_k = p_0 \left(\frac{\lambda}{M\mu}\right)^k \frac{M \cdot (M-1) \cdots (M-k+1) k!}{k! \, m!} m^{m-k}$$

$$\lim_{M \to \infty} p_k = p_0 \left(\frac{\lambda}{\mu}\right)^k \frac{m^{m-k}}{m!} = p_0 \rho^k \frac{m^m}{m!} \quad \text{(as in Eq. (3.37))}$$

(iv) M/M/1/K (section 3.6)

$$m = 1, \quad K = K, \quad M = \infty, \quad \lambda' = \lambda/M$$

Eq. (3.52) gives

$$p_k = p_0 \left(\frac{\lambda}{M\mu}\right)^k \frac{M \cdot (M-1) \cdots (M-k+1)}{k!} k! \quad 1 \leq k \leq K$$

3.8.

$$\lim_{M\to\infty} p_k = p_0\left(\frac{\lambda}{\mu}\right)^k \quad \text{for } k \leq K \quad \text{(as in Eq. (3.41))}$$

(v)   M/M/m/m   (section 3.7)

$$m = m, \quad K = m, \quad M = \infty, \quad \lambda' = \lambda/M$$

Eq. (3.51) and Eq. (3.52) give

$$p_k = p_0\left(\frac{\lambda}{M\mu}\right)^k \frac{M\cdot(M-1)\cdots(M-k+1)}{k!} \quad 0 \leq k \leq m$$

$$\lim_{M\to\infty} p_k = p_0\left(\frac{\lambda}{\mu}\right)^k \frac{1}{k!} \quad 0 \leq k \leq m \quad \text{(as in Eq. (3.45))}$$

(vi)   M/M/1//M   (section 3.8)

$$m = 1, \quad K = M, \quad M = M, \quad \lambda' = \lambda$$

Eq. (3.52) gives

$$p_k = p_0\left(\frac{\lambda}{\mu}\right)^k \frac{M!}{(M-k)!\,k!}\, k!$$

$$p_k = p_0\left(\frac{\lambda}{\mu}\right)^k \frac{M!}{(M-k)!} \quad 0 \leq k \leq M \quad \text{(as in Eq. (3.47))}$$

(vii)   M/M/∞//M   (section 3.9)

$$m = M, \quad K = M, \quad M = M, \quad \lambda' = \lambda$$

Eq. (3.51) and Eq. (3.52) give

$$p_k = p_0\left(\frac{\lambda}{\mu}\right)^k \binom{M}{k} \quad 0 \leq k \leq M \quad \text{(as in Eq. (3.49))}$$

PROBLEM 3.9.†

Using the definition for $B(m, \lambda/\mu)$ in Section 3.7 and the definition of $C(m, \lambda/\mu)$ given in Section 3.5 establish the following for $\lambda/\mu > 0$, $m = 1, 2, \ldots$

(a) $$B\left(m, \frac{\lambda}{\mu}\right) < \sum_{k=m}^{\infty} \frac{(\lambda/\mu)^k}{k!} e^{-\lambda/\mu} < C\left(m, \frac{\lambda}{\mu}\right)$$

(b) $C\left(m, \dfrac{\lambda}{\mu}\right) = \dfrac{B\left(m, \dfrac{\lambda}{\mu}\right)}{1 - \dfrac{\lambda}{m\mu}\left[1 - B\left(m, \dfrac{\lambda}{\mu}\right)\right]}$

(c) $B\left(m+1, \dfrac{\lambda}{\mu}\right) = \dfrac{\dfrac{\lambda}{\mu}B\left(m, \dfrac{\lambda}{\mu}\right)}{m+1 + \dfrac{\lambda}{\mu}B\left(m, \dfrac{\lambda}{\mu}\right)}$

SOLUTION

(a) We begin with the obvious inequality:

$$\frac{(\lambda/\mu)^m}{m!} < \sum_{k=m}^{\infty} \frac{(\lambda/\mu)^k}{k!}$$

$$\therefore \frac{(\lambda/\mu)^m}{m!} \left[\sum_{k=0}^{m-1} \frac{(\lambda/\mu)^k}{k!}\right] < \left[\sum_{k=m}^{\infty} \frac{(\lambda/\mu)^k}{k!}\right]\left[\sum_{k=0}^{m-1} \frac{(\lambda/\mu)^k}{k!}\right]$$

$$\frac{(\lambda/\mu)^m}{m!}\left[e^{\lambda/\mu} - \sum_{k=m}^{\infty} \frac{(\lambda/\mu)^k}{k!}\right] < \left[\sum_{k=m}^{\infty} \frac{(\lambda/\mu)^k}{k!}\right]\left[\sum_{k=0}^{m-1} \frac{(\lambda/\mu)^k}{k!}\right]$$

$$e^{\lambda/\mu}\frac{(\lambda/\mu)^m}{m!} < \left[\sum_{k=m}^{\infty} \frac{(\lambda/\mu)^k}{k!}\right]\left[\sum_{k=0}^{m} \frac{(\lambda/\mu)^k}{k!}\right]$$

$$\frac{(\lambda/\mu)^m/m!}{\sum_{k=0}^{m}(\lambda/\mu)^k/k!} = B\left(m, \frac{\lambda}{\mu}\right) < \sum_{k=m}^{\infty} \frac{(\lambda/\mu)^k}{k!} e^{-\lambda/\mu}$$

This establishes the left-hand inequality of (a). For the right-hand inequality, we proceed as follows: For $k > m$,

$$k! = m!\,(m+1)\cdots(k) > m!\,m^{k-m}$$

So

$$\frac{1}{k!} < \frac{1}{m!}\frac{1}{m^{k-m}}$$

$$\sum_{k=m}^{\infty} \frac{(\lambda/\mu)^k}{k!} < \sum_{k=m}^{\infty} \frac{(\lambda/\mu)^k}{m!}\frac{1}{m^{k-m}}$$

3.9.

Letting $\rho = \dfrac{\lambda}{m\mu}$, and shifting the index of summation, we have (for $\rho < 1$):

$$\sum_{k=m}^{\infty} \frac{(\lambda/\mu)^k}{k!} < \frac{(\lambda/\mu)^m}{m!} \sum_{j=0}^{\infty} \rho^j = \left[\frac{(m\rho)^m}{m!}\right]\left(\frac{1}{1-\rho}\right)$$

We now proceed as for the left-hand inequality.

$$\left[\sum_{k=m}^{\infty} \frac{(\lambda/\mu)^k}{k!}\right]\left[\sum_{k=0}^{m-1} \frac{(\lambda/\mu)^k}{k!}\right] < \left[\frac{(m\rho)^m}{m!}\right]\left(\frac{1}{1-\rho}\right)\left[e^{\lambda/\mu} - \sum_{k=m}^{\infty} \frac{(\lambda/\mu)^k}{k!}\right]$$

$$\therefore \sum_{k=m}^{\infty} \frac{(\lambda/\mu)^k}{k!} e^{-\lambda/\mu} < \frac{\left[\dfrac{(m\rho)^m}{m!}\right]\left(\dfrac{1}{1-\rho}\right)}{\left[\displaystyle\sum_{k=0}^{m-1} \frac{(\lambda/\mu)^k}{k!} + \left[\dfrac{(m\rho)^m}{m!}\right]\left(\dfrac{1}{1-\rho}\right)\right]} = C\!\left(m, \frac{\lambda}{\mu}\right)$$

(b) Since

$$1 - \frac{\lambda}{m\mu}\left[1 - B\!\left(m, \frac{\lambda}{\mu}\right)\right] = \frac{\left[1 - \dfrac{\lambda}{m\mu}\right]\left[\displaystyle\sum_{k=0}^{m-1} \frac{(\lambda/\mu)^k}{k!}\right] + \dfrac{(\lambda/\mu)^m}{m!}}{\displaystyle\sum_{k=0}^{m} \frac{(\lambda/\mu)^k}{k!}}$$

we have

$$\frac{B\!\left(m, \dfrac{\lambda}{\mu}\right)}{1 - \dfrac{\lambda}{m\mu}\left[1 - B\!\left(m, \dfrac{\lambda}{\mu}\right)\right]} = \frac{\dfrac{(\lambda/\mu)^m}{m!}}{\left[1 - \dfrac{\lambda}{m\mu}\right]\left[\displaystyle\sum_{k=0}^{m-1} \frac{(\lambda/\mu)^k}{k!}\right] + \dfrac{(\lambda/\mu)^m}{m!}}$$

$$= \frac{\left[\dfrac{(m\rho)^m}{m!}\right]\left(\dfrac{1}{1-\rho}\right)}{\left[\displaystyle\sum_{k=0}^{m-1} \frac{(\lambda/\mu)^k}{k!} + \left[\dfrac{(m\rho)^m}{m!}\right]\left(\dfrac{1}{1-\rho}\right)\right]}$$

$$= C\!\left(m, \frac{\lambda}{\mu}\right)$$

3.9.

(c)
$$B\left(m+1,\frac{\lambda}{\mu}\right) = \frac{\dfrac{(\lambda/\mu)^{m+1}}{(m+1)!}}{\sum\limits_{k=0}^{m+1}\dfrac{(\lambda/\mu)^k}{k!}} = \frac{\left[\dfrac{(\lambda/\mu)^m}{m!}\right]\left(\dfrac{\lambda}{\mu}\right)}{(m+1)\sum\limits_{k=0}^{m}\dfrac{(\lambda/\mu)^k}{k!} + \left[\dfrac{(\lambda/\mu)^m}{m!}\right]\left(\dfrac{\lambda}{\mu}\right)}$$

$$= \frac{\dfrac{\lambda}{\mu}B\left(m,\dfrac{\lambda}{\mu}\right)}{m+1+\dfrac{\lambda}{\mu}B\left(m,\dfrac{\lambda}{\mu}\right)}$$

PROBLEM 3.10.

Here we consider an M/M/1 queue in discrete time where time is segmented into intervals of length q sec each. We assume that events can only occur at the ends of these discrete time intervals. In particular the probability of a single arrival at the end of such an interval is given by λq and the probability of no arrival at that point is $1-\lambda q$ (thus at most one arrival may occur). Similarly the departure process is such that if a customer is in service during an interval he will complete service at the end of that interval with probability $1-\sigma$ or will require at least one more interval with probability σ.

(a) Derive the form for $a(t)$ and $b(x)$, the interarrival time and service time pdf's, respectively.
(b) Assuming FCFS, write down the equilibrium equations that govern the behavior of $p_k = P[k$ customers in system at the end of a discrete time interval] where k includes any arrivals who have occurred at the end of this interval as well as any customers who are about to leave at this point.
(c) Solve for the expected value of the number of customers at these points.

SOLUTION

(a) In order that the interarrival time $\tilde{t} = nq$, we require no arrivals for $n-1$ consecutive intervals, followed by an arrival.
$$P[\tilde{t} = nq] = \lambda q(1-\lambda q)^{n-1} \quad n = 1, 2, \ldots$$

Similarly
$$P[\tilde{x} = nq] = (1-\sigma)\sigma^{n-1} \quad n = 1, 2, \ldots$$
and so
$$a(t) = \sum_{n=1}^{\infty} \lambda q (1-\lambda q)^{n-1} u_0(t-nq) \qquad \blacksquare$$

$$b(x) = \sum_{n=1}^{\infty} (1-\sigma)\sigma^{n-1} u_0(x-nq) \qquad \blacksquare$$

(b) The number in system is a discrete-state discrete-time Markov chain whose equilibrium equations are:

$$\lambda q p_0 = (1-\lambda q)(1-\sigma) p_1 \quad k = 0 \qquad \blacksquare$$

$$[\lambda q \sigma + (1-\lambda q)(1-\sigma)] p_1 = \lambda q p_0 + (1-\lambda q)(1-\sigma) p_2 \quad k = 1 \qquad \blacksquare$$

$$[\lambda q \sigma + (1-\lambda q)(1-\sigma)] p_k = \lambda q \sigma p_{k-1} + (1-\lambda q)(1-\sigma) p_{k+1} \quad k \geq 2 \qquad \blacksquare$$

(c) First we solve for p_k by iteration. We find $\rho = \bar{x}/\bar{t}$ as follows:

$$\bar{t} = \sum_{n=1}^{\infty} (nq) \lambda q (1-\lambda q)^{n-1} = \frac{1}{\lambda}$$

$$\bar{x} = \sum_{n=1}^{\infty} (nq)(1-\sigma)\sigma^{n-1} = \frac{q}{1-\sigma}$$

and thus
$$\rho = \frac{\lambda q}{1-\sigma}$$

Now for the iteration. We have
$$p_1 = \frac{\rho}{1-\lambda q} p_0$$

$$p_2 = \frac{\rho^2 \sigma}{(1-\lambda q)^2} p_0$$

and since
$$p_k = \frac{\lambda q \sigma}{(1-\lambda q)(1-\sigma)} p_{k-1} \quad k \geq 3$$

our iteration yields
$$p_k = \left[\frac{\rho \sigma}{1-\lambda q}\right]^k \frac{1}{\sigma} p_0 \quad \text{for } k \geq 1$$

3.10.

[Note: for stability, $\dfrac{\rho\sigma}{1-\lambda q} < 1$, that is $\lambda q\sigma < (1-\lambda q)(1-\sigma)$. Thus $\lambda q\sigma < 1-\lambda q-\sigma+\lambda q\sigma$ or $\lambda q < 1-\sigma \Rightarrow \rho < 1$.]

To find p_0,

$$\sum_{k=0}^{\infty} p_k = 1 \quad \Rightarrow \quad p_0 + \sum_{k=1}^{\infty}\left[\dfrac{\rho\sigma}{1-\lambda q}\right]^k \dfrac{1}{\sigma} p_0 = 1$$

$$p_0\left\{1+\dfrac{1}{\sigma}\left[\dfrac{1}{1-\dfrac{\rho\sigma}{1-\lambda q}}-1\right]\right\} = 1$$

$$p_0\left\{1+\dfrac{1}{\sigma}\dfrac{\rho\sigma}{1-\lambda q-\rho\sigma}\right\} = 1$$

$$\therefore\ p_0 = 1-\rho \qquad \blacksquare$$

$$p_k = \dfrac{1-\rho}{\sigma}\left[\dfrac{\rho\sigma}{1-\lambda q}\right]^k \quad k \geqslant 1 \qquad \blacksquare$$

Finally, the mean number in system at these points in time is

$$\overline{N} = \sum_{k=1}^{\infty} k\, p_k = \dfrac{1-\rho}{\sigma}\dfrac{\rho\sigma}{1-\lambda q}\sum_{k=1}^{\infty} k\left[\dfrac{\rho\sigma}{1-\lambda q}\right]^{k-1}$$

$$= \dfrac{(1-\rho)\rho}{1-\lambda q}\dfrac{1}{\left(1-\dfrac{\rho\sigma}{1-\lambda q}\right)^2}$$

$$\overline{N} = \dfrac{\rho}{1-\rho}(1-\lambda q) \qquad \blacksquare$$

[$\overline{N}$ can also be found from the z-transform $P(z) = \sum_{k=0}^{\infty} p_k z^k$ which turns out to be

$$P(z) = (1-\rho)\left[1+\dfrac{\lambda qz}{(1-\lambda q)(1-\sigma)-\lambda q\sigma z}\right]$$

Then $\overline{N} = P^{(1)}(1)$.]

3.10.

PROBLEM 3.11.

Consider an M/M/1 system with "feedback"; by this we mean that when a customer departs from service he has probability σ of rejoining the tail of the queue after a random feedback time, which is exponentially distributed (with mean $1/\gamma$ sec); on the other hand, with probability $1-\sigma$ he will depart forever after completing service. It is clear that a customer may return many times to the tail of the queue before making his eventual final departure. Let p_{kj} be the equilibrium probability that there are k customers in the "system" (that is, in the queue and the service facility) and that there are j customers in the process of returning to the system.

(a) Write down the set of difference equations for the equilibrium probabilities p_{kj}.
(b) Defining the double z-transform

$$P(z_1, z_2) = \sum_{k=0}^{\infty} \sum_{j=0}^{\infty} p_{kj} z_1^k z_2^j$$

show that

$$\gamma(z_2 - z_1) \frac{\partial P(z_1, z_2)}{\partial z_2} + \left\{ \lambda(1 - z_1) + \mu \left[1 - \frac{1-\sigma}{z_1} - \sigma \frac{z_2}{z_1} \right] \right\} P(z_1, z_2)$$

$$= \mu \left[1 - \frac{1-\sigma}{z_1} - \sigma \frac{z_2}{z_1} \right] P(0, z_2)$$

(c) By taking advantage of the moment-generating properties of our z-transforms, show that the mean number in the "system" (queue plus server) is given by $\rho/(1-\rho)$ and that the mean number returning to the tail of the queue is given by $\mu\sigma\rho/\gamma$, where $\rho = \lambda/(1-\sigma)\mu$.

SOLUTION

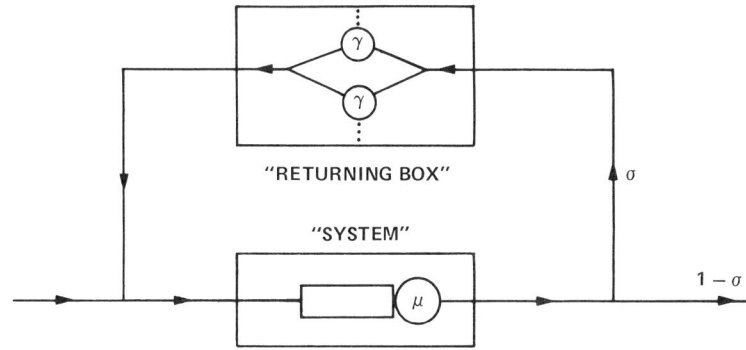

3.11.

(a) **Case (1):** $k=0,\ j=0$
$$\lambda p_{00} = \mu(1-\sigma)p_{10} \qquad \blacksquare$$

Case (2): $k=0,\ j>0$
$$(\lambda + j\gamma)p_{0j} = \mu(1-\sigma)p_{1j} + \mu\sigma p_{1,j-1} \qquad \blacksquare$$

Case (3): $k>0,\ j=0$
$$(\lambda + \mu)p_{k0} = \lambda p_{k-1,0} + \mu(1-\sigma)p_{k+1,0} + \gamma p_{k-1,1} \qquad \blacksquare$$

Case (4): $k>0,\ j>0$
$$(\lambda + \mu + j\gamma)p_{kj} = \lambda p_{k-1,j} + \mu(1-\sigma)p_{k+1,j}$$
$$+ \mu\sigma p_{k+1,j-1} + (j+1)\gamma p_{k-1,j+1} \qquad \blacksquare$$

(b) We multiply the (k,j) equation by $z_1^k z_2^j$ and sum, case by case:
Case (1) yields:
$$\lambda p_{00} = \mu(1-\sigma)p_{10}$$

Case (2) yields:
$$\sum_{j=1}^{\infty}(\lambda+j\gamma)p_{0j}z_2^j = \sum_{j=1}^{\infty}\mu(1-\sigma)p_{1j}z_2^j + \sum_{j=1}^{\infty}\mu\sigma p_{1,j-1}z_2^j$$

Case (3) yields:
$$\sum_{k=1}^{\infty}(\lambda+\mu)p_{k0}z_1^k = \sum_{k=1}^{\infty}\lambda p_{k-1,0}z_1^k + \sum_{k=1}^{\infty}\mu(1-\sigma)p_{k+1,0}z_1^k$$
$$+ \sum_{k=1}^{\infty}\gamma p_{k-1,1}z_1^k$$

Case (4) yields:
$$\sum_{k=1}^{\infty}\sum_{j=1}^{\infty}(\lambda+\mu+j\gamma)p_{kj}z_1^k z_2^j = \sum_{k=1}^{\infty}\sum_{j=1}^{\infty}\lambda p_{k-1,j}z_1^k z_2^j$$
$$+ \sum_{k=1}^{\infty}\sum_{j=1}^{\infty}\mu(1-\sigma)p_{k+1,j}z_1^k z_2^j + \sum_{k=1}^{\infty}\sum_{j=1}^{\infty}\mu\sigma p_{k+1,j-1}z_1^k z_2^j$$
$$+ \sum_{k=1}^{\infty}\sum_{j=1}^{\infty}(j+1)\gamma p_{k-1,j+1}z_1^k z_2^j$$

3.11.

Summing these four equations gives:

$$\lambda P(z_1, z_2) + \mu\left[P(z_1, z_2) - \sum_{j=0}^{\infty} p_{0j} z_2^j\right] + \gamma z_2 \frac{\partial}{\partial z_2} P(z_1, z_2) =$$

$$\lambda z_1 P(z_1, z_2) + \frac{\mu(1-\sigma)}{z_1}\left[P(z_1, z_2) - \sum_{j=0}^{\infty} p_{0j} z_2^j\right]$$

$$+ \mu\sigma \frac{z_2}{z_1}\left[P(z_1, z_2) - \sum_{j=0}^{\infty} p_{0j} z_2^j\right] + \gamma z_1 \frac{\partial}{\partial z_2} P(z_1, z_2)$$

Noting that

$$P(z_1, z_2) = \sum_{j=0}^{\infty} p_{0j} z_2^j + \sum_{k=1}^{\infty} \sum_{j=0}^{\infty} p_{kj} z_1^k z_2^j$$

and thus

$$P(0, z_2) = \sum_{j=0}^{\infty} p_{0j} z_2^j$$

we finally have the required equation:

$$\gamma(z_2 - z_1)\frac{\partial P(z_1, z_2)}{\partial z_2} + \left\{\lambda(1 - z_1) + \mu\left[1 - \frac{1-\sigma}{z_1} - \sigma\frac{z_2}{z_1}\right]\right\} P(z_1, z_2)$$

$$= \mu\left[1 - \frac{1-\sigma}{z_1} - \sigma\frac{z_2}{z_1}\right] P(0, z_2)$$

(c) The mean number of customers in the "returning box" is

$$\bar{N}_2 = \sum_{j=1}^{\infty} j\left(\sum_{k=0}^{\infty} p_{kj}\right)$$

which may also be written as

$$\bar{N}_2 = \left.\frac{\partial}{\partial z_2} P(z_1, z_2)\right|_{z_1 = z_2 = 1}$$

From this, we find $\bar{N}_2$ as follows. First, substitute $z_1 = 1$ in the equation derived in part (b) to give

$$\gamma(z_2 - 1)\frac{\partial}{\partial z_2} P(1, z_2) + \mu\sigma(1 - z_2) P(1, z_2) = \mu\sigma(1 - z_2) P(0, z_2)$$

Now, for $z_2 \neq 1$,

$$\gamma \frac{\partial}{\partial z_2} P(1, z_2) = \mu\sigma[P(1, z_2) - P(0, z_2)]$$

3.11.

Letting $z_2 \to 1$ and using analyticity we obtain
$$\gamma \bar{N}_2 = \mu\sigma[P(1,1) - P(0,1)]$$
Clearly $P(1,1) = 1$. To find $P(0,1)$, we put $z_1 = z_2 = z$ in the equation from part (b):
$$\left[\lambda(1-z) + \mu(1-\sigma)\left(1 - \frac{1}{z}\right)\right]P(z,z) = \mu(1-\sigma)\left(1 - \frac{1}{z}\right)P(0,z)$$
or
$$P(z,z)[\mu(1-\sigma) - \lambda z] = \mu(1-\sigma)P(0,z)$$
Letting $z \to 1$, we get
$$P(1,1)[\mu(1-\sigma) - \lambda] = \mu(1-\sigma)P(0,1)$$
and so
$$P(0,1) = 1 - \frac{\lambda}{\mu(1-\sigma)} = 1 - \rho$$
$$\therefore \bar{N}_2 = \frac{\mu\sigma\rho}{\gamma} \qquad \blacksquare$$

[Note: the arrival rate, say η, to the "system" satisfies $\eta = \lambda + \eta\sigma$ or $\eta = \frac{\lambda}{1-\sigma}$. Therefore $\rho = \frac{\eta}{\mu} = \frac{\lambda}{\mu(1-\sigma)}$.]

We observe that the mean number of customers in the "system" is
$$\bar{N}_1 = \sum_{k=1}^{\infty} k \left(\sum_{j=0}^{\infty} p_{kj}\right)$$
In order to find $\bar{N}_1$, we first define
$$Q(z) = P(z,z) = \sum_{k=0}^{\infty} \sum_{j=0}^{\infty} p_{kj} z^{k+j}$$
Therefore,
$$\left.\frac{dQ(z)}{dz}\right|_{z=1} = \sum_{k=0}^{\infty} \sum_{j=0}^{\infty} (k+j) p_{kj} = \bar{N}_1 + \bar{N}_2$$
We already know that
$$P(z,z)[\mu(1-\sigma) - \lambda z] = \mu(1-\sigma)P(0,z)$$
and so
$$Q(z)(1 - \rho z) = P(0,z)$$

3.11.

Differentiating with respect to z gives

$$\frac{dQ(z)}{dz}(1-\rho z) - \rho Q(z) = \frac{\partial}{\partial z}P(0,z)$$

Letting $z \to 1$ (use $Q(1) = 1$) yields

$$\left[\bar{N}_1 + \bar{N}_2\right](1-\rho) = \rho + \left.\frac{\partial P(0,z)}{\partial z}\right|_{z=1}$$

$$\bar{N}_1 + \bar{N}_2 = \frac{\rho}{1-\rho} + \left(\frac{1}{1-\rho}\right)\left.\frac{\partial P(0,z)}{\partial z}\right|_{z=1}$$

Now

$$\left(\frac{1}{1-\rho}\right)\left.\frac{\partial P(0,z)}{\partial z}\right|_{z=1} = \frac{\left.\frac{\partial P(0,z)}{\partial z}\right|_{z=1}}{P(0,1)}$$

and from the expression we derived in (b) for $P(0, z_2)$, we see that this represents the mean number of customers in the "returning box" conditioned on there being no one in the "system". Assuming this is independent of the state of the "system", it is just $\bar{N}_2$. Thus

$$\bar{N}_1 = \frac{\rho}{1-\rho} \qquad \blacksquare$$

PROBLEM 3.12.

Consider a "cyclic queue" in which M customers circulate around through two queueing facilities as shown below.

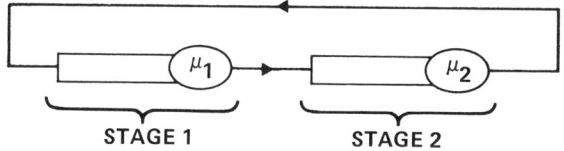

Both servers are of the exponential type with rates μ_1 and μ_2, respectively. Let

$$p_k = P[k \text{ customers in stage 1 and } M-k \text{ in stage 2}]$$

(a) Draw the state-transition-rate diagram.
(b) Write down the relationship among $\{p_k\}$.

(c) Find
$$P(z) = \sum_{k=0}^{M} p_k z^k$$

(d) Find p_k.

SOLUTION

(a)

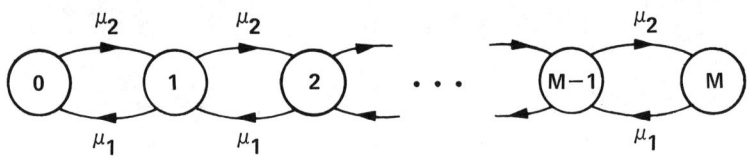

[We note that this is exactly the state-transition-rate diagram of the finite storage system of Section 3.6 (with $K = M$). Therefore we need proceed no further to find p_k. However, in this problem, we seek the balance equations and the z-transform $P(z)$.]

(b)
$$\mu_2 p_0 = \mu_1 p_1 \quad k = 0 \qquad \blacksquare$$

$$(\mu_1 + \mu_2) p_k = \mu_2 p_{k-1} + \mu_1 p_{k+1} \quad 1 \leq k \leq M-1 \qquad \blacksquare$$

$$\mu_1 p_M = \mu_2 p_{M-1} \quad k = M \qquad \blacksquare$$

(c) Multiplying the kth equation ($0 \leq k \leq M$) by z^k and summing,

$$\mu_1 \sum_{k=1}^{M} p_k z^k + \mu_2 \sum_{k=0}^{M-1} p_k z^k = \mu_1 \sum_{k=0}^{M-1} p_{k+1} z^k + \mu_2 \sum_{k=1}^{M} p_{k-1} z^k$$

and so

$$\mu_1 [P(z) - p_0] + \mu_2 [P(z) - p_M z^M] = \frac{\mu_1}{z}[P(z) - p_0] + \mu_2 z [P(z) - p_M z^M]$$

$$\therefore \; P(z) = \frac{p_0 - \dfrac{\mu_2}{\mu_1} z p_M z^M}{1 - \dfrac{\mu_2}{\mu_1} z}$$

3.12.

Since $P(z)$ is analytic everywhere in the finite z-plane, the root $z = \dfrac{\mu_1}{\mu_2}$ of the denominator must also be a root of the numerator. So

$$p_0 - \frac{\mu_2}{\mu_1}\left(\frac{\mu_1}{\mu_2}\right)p_M\left(\frac{\mu_1}{\mu_2}\right)^M = 0$$

or

$$p_M = p_0\left(\frac{\mu_2}{\mu_1}\right)^M$$

Therefore

$$P(z) = \frac{p_0 - p_0\left(\dfrac{\mu_2}{\mu_1}\right)^{M+1} z^{M+1}}{1 - \dfrac{\mu_2}{\mu_1} z}$$

or

$$P(z) = p_0 \sum_{k=0}^{M} \left(\frac{\mu_2}{\mu_1} z\right)^k$$

Now for p_0:

$$1 = P(1) = p_0 \sum_{k=0}^{M} \left(\frac{\mu_2}{\mu_1}\right)^k$$

So

$$P(z) = \frac{\sum_{k=0}^{M}\left(\dfrac{\mu_2}{\mu_1}\right)^k z^k}{\sum_{k=0}^{M}\left(\dfrac{\mu_2}{\mu_1}\right)^k}\qquad\blacksquare$$

(d) By inspection of the power series in (c), we get

$$p_k = \frac{\left(\dfrac{\mu_2}{\mu_1}\right)^k}{\sum_{k=0}^{M}\left(\dfrac{\mu_2}{\mu_1}\right)^k} \qquad 0 \leqslant k \leqslant M \qquad\blacksquare$$

which may also be expressed as

3.12.

$$p_k = \frac{1 - \dfrac{\mu_2}{\mu_1}}{1 - \left(\dfrac{\mu_2}{\mu_1}\right)^{M+1}} \left(\frac{\mu_2}{\mu_1}\right)^k \quad 0 \leqslant k \leqslant M \qquad \blacksquare$$

a result identical to Eq. (3.43).

PROBLEM 3.13.

Consider an M/M/1 queue with parameters λ and μ. A customer in the queue will defect (depart without service) with probability $\alpha \Delta t + o(\Delta t)$ in any interval of duration Δt.
(a) Draw the state-transition-rate diagram.
(b) Express p_{k+1} in terms of p_k.
(c) For $\alpha = \mu$, solve for p_k ($k = 0, 1, 2, \ldots$).

SOLUTION

(a)

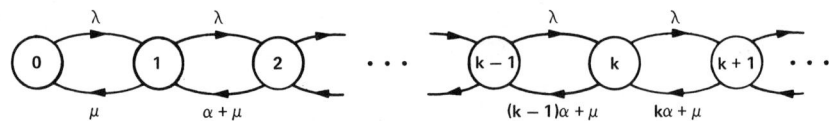

(b)

$$\lambda p_0 = \mu p_1 \quad k = 0$$

$$[\lambda + (k-1)\alpha + \mu] p_k = \lambda p_{k-1} + (k\alpha + \mu) p_{k+1} \quad k \geqslant 1$$

By recursion we find

$$(k\alpha + \mu) p_{k+1} = \lambda p_k \qquad \blacksquare$$

(This could also have been found directly by using Eq. (3.7).)

(c) First observe from part (b) that, for $k \geqslant 1$, we have $\mu_k = (k-1)\alpha + \mu$. Thus $\mu_k = k\mu$ for $\alpha = \mu$. Note that λ_k and μ_k for this system ($\alpha = \mu$) are exactly the same as for M/M/∞. Therefore, the solution for p_k must be the same (although other system parameters such as waiting time and queue size will be different). Then Eq. (3.34) gives

$$p_k = \frac{\left(\frac{\lambda}{\mu}\right)^k}{k!} e^{-\frac{\lambda}{\mu}} \quad k = 0, 1, 2, \ldots \quad \blacksquare$$

PROBLEM 3.14.

Let us elaborate on the M/M/1/K system of Section 3.6.
(a) Evaluate p_k when $\lambda = \mu$.
(b) Find $\overline{N}$ for $\lambda \neq \mu$ and for $\lambda = \mu$.
(c) Find T by carefully solving for the average arrival rate to the system.

SOLUTION

(a) Eq. (3.41) and Eq. (3.42) give

$$p_k = p_0 \left(\frac{\lambda}{\mu}\right)^k \quad k \leqslant K$$
$$p_k = 0 \quad k > K$$

Thus for $\lambda = \mu$ we find

$$p_k = p_0 \quad \text{for } k \leqslant K$$

Since $\sum_{k=0}^{K} p_k = 1$, then

$$p_k = \frac{1}{K+1} \quad 0 \leqslant k \leqslant K \quad \blacksquare$$

(b) (i) For $\lambda = \mu$,

$$\overline{N} = \sum_{k=1}^{K} k p_k = \sum_{k=1}^{K} k \frac{1}{K+1} = \frac{1}{K+1} \sum_{k=1}^{K} k = \frac{1}{K+1} \cdot \frac{K(K+1)}{2}$$

$$\overline{N} = \frac{K}{2} \quad \blacksquare$$

(ii) For $\lambda \neq \mu$,

$$\bar{N} = \sum_{k=1}^{K} k p_k = \sum_{k=1}^{K} k p_0 \left(\frac{\lambda}{\mu}\right)^k = p_0 \left(\frac{\lambda}{\mu}\right) \frac{d}{d\left(\frac{\lambda}{\mu}\right)} \left[\sum_{k=0}^{K} \left(\frac{\lambda}{\mu}\right)^k\right]$$

Recall that $p_0 = \dfrac{1 - \dfrac{\lambda}{\mu}}{1 - \left(\dfrac{\lambda}{\mu}\right)^{K+1}}$ by Eq. (3.43). So

$$\bar{N} = \frac{1 - \dfrac{\lambda}{\mu}}{1 - \left(\dfrac{\lambda}{\mu}\right)^{K+1}} \left(\frac{\lambda}{\mu}\right) \left[\frac{1 - (K+1)\left(\dfrac{\lambda}{\mu}\right)^K + K\left(\dfrac{\lambda}{\mu}\right)^{K+1}}{\left(1 - \dfrac{\lambda}{\mu}\right)^2}\right]$$

or

$$\bar{N} = \left(\frac{\lambda}{\mu}\right) \left[\frac{1 - (K+1)\left(\dfrac{\lambda}{\mu}\right)^K + K\left(\dfrac{\lambda}{\mu}\right)^{K+1}}{\left[1 - \left(\dfrac{\lambda}{\mu}\right)^{K+1}\right]\left(1 - \dfrac{\lambda}{\mu}\right)}\right] \quad \blacksquare$$

(c) (i) For $\lambda = \mu$, $p_k = \dfrac{1}{K+1}$ $0 \leq k \leq K$ from part (a). The average arrival rate $\bar{\lambda}$ is

$$\bar{\lambda} = \sum_{k=0}^{K} \lambda_k p_k = \lambda \sum_{k=0}^{K-1} \frac{1}{K+1} \quad (\lambda_K = 0)$$

$$\bar{\lambda} = \lambda \frac{K}{K+1} \quad \blacksquare$$

By Little's result, and also part (b)

$$T = \frac{\bar{N}}{\bar{\lambda}} = \frac{K/2}{\lambda K/(K+1)} = \frac{K+1}{2\lambda} \quad \blacksquare$$

(ii) For $\lambda \neq \mu$, from Eq. (3.43)

$$\bar{\lambda} = \sum_{k=0}^{K} \lambda_k p_k = \lambda \sum_{k=0}^{K-1} \frac{1 - \dfrac{\lambda}{\mu}}{1 - \left(\dfrac{\lambda}{\mu}\right)^{K+1}} \left(\frac{\lambda}{\mu}\right)^k$$

3.14.

$$\bar{\lambda} = \lambda \frac{1 - \left(\frac{\lambda}{\mu}\right)^{K}}{1 - \left(\frac{\lambda}{\mu}\right)^{K+1}}$$ ∎

Little's result and part (b) give

$$T = \frac{\bar{N}}{\bar{\lambda}} = \left(\frac{\lambda}{\mu}\right) \left[\frac{1 - (K+1)\left(\frac{\lambda}{\mu}\right)^{K} + K\left(\frac{\lambda}{\mu}\right)^{K+1}}{\left(1 - \left(\frac{\lambda}{\mu}\right)^{K+1}\right)\left(1 - \frac{\lambda}{\mu}\right)} \right] \left[\frac{1 - \left(\frac{\lambda}{\mu}\right)^{K+1}}{\lambda\left(1 - \left(\frac{\lambda}{\mu}\right)^{K}\right)} \right]$$

$$T = \frac{1}{\mu} \left[\frac{1 - (K+1)\left(\frac{\lambda}{\mu}\right)^{K} + K\left(\frac{\lambda}{\mu}\right)^{K+1}}{\left(1 - \left(\frac{\lambda}{\mu}\right)^{K}\right)\left(1 - \frac{\lambda}{\mu}\right)} \right]$$ ∎

3.14.

Chapter 4
Markovian Queues in Equilibrium

PROBLEM 4.1.

Consider the Markovian queueing system shown below. Branch labels are birth and death rates. Node labels give the number of customers in the system.

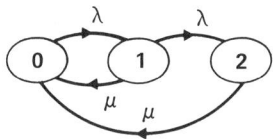

(a) Solve for p_k.
(b) Find the average number in the system.
(c) For $\lambda = \mu$, what values do we get for parts (a) and (b)? Try to interpret these results.
(d) Write down the transition rate matrix $\mathbf{Q}$ for this problem and give the matrix equation relating $\mathbf{Q}$ to the probabilities found in part (a).

SOLUTION

(a) Using the flow conservation law for states 0 and 2 and the conservation of probability, we get the following three independent equations:

$$\lambda p_0 = \mu p_1 + \mu p_2$$

$$\mu p_2 = \lambda p_1$$

$$p_0 + p_1 + p_2 = 1$$

Solving this gives

$$p_0 = \frac{\mu}{\lambda + \mu} \qquad \blacksquare$$

$$p_1 = \frac{\lambda \mu}{(\lambda + \mu)^2} \qquad \blacksquare$$

4.1.

$$p_2 = \frac{\lambda^2}{(\lambda+\mu)^2}$$

(b)

$$\bar{N} = 0 \cdot p_0 + 1 \cdot p_1 + 2 \cdot p_2 = \frac{\lambda\mu + 2\lambda^2}{(\lambda+\mu)^2}$$

$$\bar{N} = \frac{\lambda(2\lambda+\mu)}{(\lambda+\mu)^2}$$

(c) If $\lambda = \mu$, the results in parts (a) and (b) become

$$p_0 = \frac{1}{2}, \quad p_1 = p_2 = \frac{1}{4}, \quad \bar{N} = \frac{3}{4}$$

To interpret these results, consider a cycle from E_0 back to E_0. The rate out of E_0 is λ $(= \mu)$ which puts the system into E_1. The rate out of E_1 is $\lambda + \mu = 2\mu$ and so the fraction of time spent in E_1 must be half that spent in E_0. From E_1 we arrive at E_2 with probability 1/2 (or return directly to E_0 with probability 1/2) and depart E_2 at rate μ; therefore, we spend as much time, on the average, in E_2 (i.e. $\frac{1}{2} \cdot \frac{1}{\mu}$) as in E_1 (i.e. $\frac{1}{2\mu}$).

(d) Recall that $-q_{ii}$ is the rate at which the system departs from state E_i, while q_{ij} $(i \neq j)$ is the rate at which it moves from state E_i to state E_j. Thus

$$\mathbf{Q} = \begin{bmatrix} -\lambda & \lambda & 0 \\ \mu & -(\mu+\lambda) & \lambda \\ \mu & 0 & -\mu \end{bmatrix}$$

Also recall from Eq. (2.116) that

$$\boldsymbol{\pi}\mathbf{Q} = 0 \quad (\boldsymbol{\pi} = \mathbf{p} = [p_0, p_1, p_2])$$

PROBLEM 4.2.†

Consider an $E_k/E_n/1$ queueing system where *no* queue is permitted to form. A customer who arrives to find the service facility busy is "lost" (he departs with no service). Let E_{ij} be the system state in which the "arriving" customer is in the ith arrival stage and the customer in service is in the jth service stage (note that there is always some customer in the arrival mechanism and that if there is no customer in the service facility, then we let $j=0$). Let $1/k\lambda$ be the average time spent in any arrival stage and $1/n\mu$ be the average time spent in any service stage.

4.1.−4.2.

(a) Draw the state transition diagram showing all the transition *rates*.
(b) Write down the equilibrium equation for E_{ij} where $1 < i < k$, $1 < j \leqslant n$.

SOLUTION

(a)

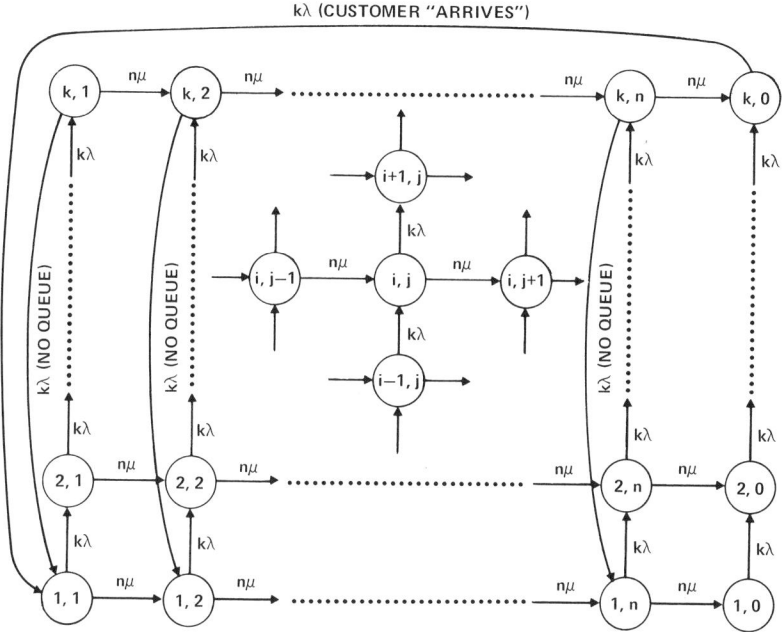

(b)

$$\text{Flow out} = \text{Flow in}$$

$$(k\lambda + n\mu)p_{ij} = k\lambda p_{i-1,j} + n\mu p_{i,j-1} \quad \text{for} \quad 1 < i < k, 1 < j \leqslant n \quad \blacksquare$$

4.2.

PROBLEM 4.3.

Consider an M/E$_r$/1 system in which *no* queue is allowed to form. Let j = the number of stages of service left in the system and let P_j be the equilibrium probability of being in state E_j.
(a) Find P_j, $j = 0, 1, \ldots, r$.
(b) Find the probability of a busy system.

SOLUTION

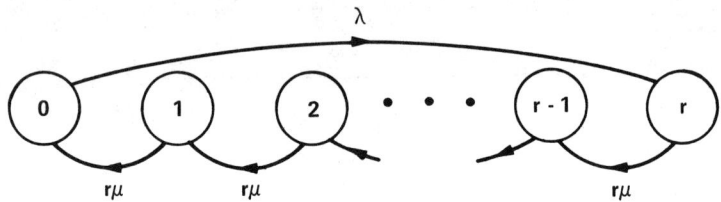

(a) The flow equations are:
$$\lambda P_0 = r\mu P_1 \qquad j=0$$
$$r\mu P_j = r\mu P_{j+1} \qquad 1 \leqslant j \leqslant r-1$$
$$r\mu P_r = \lambda P_0 \qquad j=r$$

Of these $r+1$ equations, one is redundant; using the first r we see that
$$\frac{\lambda}{r\mu} P_0 = P_1 = P_2 = \cdots = P_{r-1} = P_r$$

Also
$$\sum_{j=0}^{r} P_j = 1 \quad \Rightarrow \quad P_0 + \sum_{j=1}^{r} \frac{\lambda}{r\mu} P_0 = 1$$

Thus
$$P_0 = \frac{\mu}{\lambda + \mu} \qquad \blacksquare$$

and therefore
$$P_j = \frac{\lambda}{r(\lambda + \mu)} \qquad 1 \leqslant j \leqslant r \qquad \blacksquare$$

(b)

$$P[\text{busy system}] = 1 - P_0 = 1 - \frac{\mu}{\lambda + \mu}$$

$$P[\text{busy system}] = \frac{\lambda}{\lambda + \mu}$$ ∎

PROBLEM 4.4.

Consider an M/H$_2$/1 system in which *no* queue is allowed to form. Service is of the hyperexponential type as shown in Figure 4.10 with $\mu_1 = 2\mu\alpha_1$ and $\mu_2 = 2\mu(1-\alpha_1)$.
(a) Solve for the equilibrium probability of an empty system.
(b) Find the probability that server 1 is occupied.
(c) Find the probability of a busy system.

SOLUTION

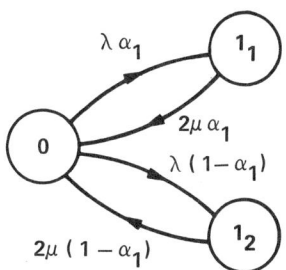

As usual, we have two independent flow equations and the conservation of probability:

$$\lambda p_0 = 2\mu\alpha_1 p_{1_1} + 2\mu(1-\alpha_1) p_{1_2}$$

$$\lambda \alpha_1 p_0 = 2\mu\alpha_1 p_{1_1}$$

$$p_0 + p_{1_1} + p_{1_2} = 1$$

Thus

$$p_0 = \frac{\mu}{\lambda+\mu}$$

$$p_{1_1} = p_{1_2} = \frac{\lambda}{2(\lambda+\mu)}$$

(a)

$$P[\text{empty system}] = p_0 = \frac{\mu}{\lambda+\mu}$$ ∎

(b)

$$P[\text{server 1 busy}] = p_{1_1} = \frac{\lambda}{2(\lambda+\mu)}$$ ∎

(c)

$$P[\text{busy system}] = 1 - p_0 = p_{1_1} + p_{1_2} = \frac{\lambda}{\lambda+\mu}$$ ∎

PROBLEM 4.5.

Consider an M/M/1 system with parameters λ and μ in which exactly two customers arrive at each arrival instant.
(a) Draw the state-transition-rate diagram.
(b) By inspection, write down the equilibrium equations for p_k ($k = 0, 1, 2, \ldots$).
(c) Let $\rho = 2\lambda/\mu$. Express $P(z)$ in terms of ρ and z.
(d) Find $P(z)$ by using the bulk arrival results from Section 4.5.
(e) Find the mean and variance of the number of customers in the system from $P(z)$.
(f) Repeat parts (a)-(e) with exactly r customers arriving at each arrival instant (and $\rho = r\lambda/\mu$).

SOLUTION

In this problem, we carry out the details of the verbal argument given in the lead paragraph of Section 4.5.

(a)

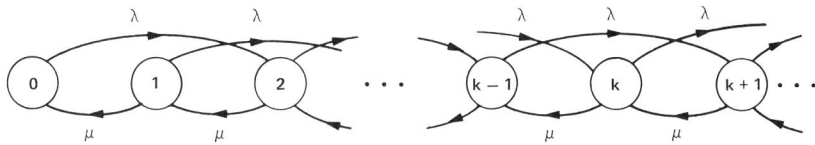

(b)

$$\lambda p_0 = \mu p_1 \quad k = 0$$

$$(\lambda + \mu) p_1 = \mu p_2 \quad k = 1$$

$$(\lambda + \mu) p_k = \lambda p_{k-2} + \mu p_{k+1} \quad k \geq 2$$

(c) Multiply the kth equation by z^k and sum for $k \geq 0$. This gives

$$\lambda \sum_{k=0}^{\infty} p_k z^k + \mu \sum_{k=1}^{\infty} p_k z^k = \lambda \sum_{k=2}^{\infty} p_{k-2} z^k + \mu \sum_{k=0}^{\infty} p_{k+1} z^k$$

$$\lambda P(z) + \mu[P(z) - p_0] = \lambda z^2 P(z) + \frac{\mu}{z}[P(z) - p_0]$$

$$P(z) = \frac{\mu p_0 \left(1 - \frac{1}{z}\right)}{\lambda(1 - z^2) + \mu\left(1 - \frac{1}{z}\right)} = \frac{\mu p_0}{\mu - \lambda z(z+1)}$$

(Note that the average arrival rate $\bar{\lambda} = 2\lambda$, and so $\rho = \bar{\lambda}\bar{x} = \frac{2\lambda}{\mu}$.) Thus

$$P(z) = \frac{2p_0}{2 - \rho z(z+1)}$$

Since $P(1) = 1 = \frac{2p_0}{2 - 2\rho}$ we have $p_0 = 1 - \rho$. Hence

$$P(z) = \frac{2(1 - \rho)}{2 - \rho z(z+1)}$$

(d) By Eq. (4.49),

$$P(z) = \frac{\mu(1 - \rho)(1 - z)}{\mu(1 - z) - \lambda z[1 - G(z)]}$$

In the system under consideration, bulks have constant size 2. Thus $G(z) = z^2$ (and $\rho = \frac{\lambda G^{(1)}(1)}{\mu} = \frac{2\lambda}{\mu}$).

4.5.

$$\therefore P(z) = \frac{\mu(1-\rho)(1-z)}{\mu(1-z) - \lambda z(1-z^2)}$$

This simplifies as before to

$$P(z) = \frac{2(1-\rho)}{2 - \rho z(z+1)} \qquad \blacksquare$$

(e) The mean and variance of the number of customers may be found from the first and second derivatives of $P(z)$. We find that

$$\frac{dP(z)}{dz} = \frac{2(1-\rho)\rho(2z+1)}{[2 - \rho z(z+1)]^2}$$

$$\overline{N} = \left.\frac{dP(z)}{dz}\right|_{z=1} = \frac{2(1-\rho)\rho(3)}{(2-2\rho)^2}$$

$$\overline{N} = \frac{3}{2} \frac{\rho}{1-\rho} \qquad \blacksquare$$

After simplification, the second derivative is

$$\frac{d^2P(z)}{dz^2} = 4(1-\rho)\rho\left[\frac{[2-\rho z(z+1)] + \rho(2z+1)^2}{[2-\rho z(z+1)]^3}\right]$$

$$\overline{N^2} - \overline{N} = \left.\frac{d^2P(z)}{dz^2}\right|_{z=1} = 4(1-\rho)\rho\left[\frac{2-2\rho+9\rho}{(2-2\rho)^3}\right]$$

$$= \frac{\rho}{2(1-\rho)^2}(2+7\rho)$$

By definition, we have,

$$\sigma_N^2 = \overline{N^2} - (\overline{N})^2 = (\overline{N^2} - \overline{N}) + \overline{N} - (\overline{N})^2$$

$$= \frac{\rho}{2(1-\rho)^2}(2+7\rho) + \frac{3}{2}\frac{\rho}{1-\rho} - \frac{9}{4}\frac{\rho^2}{(1-\rho)^2}$$

$$\sigma_N^2 = \frac{\rho(10-\rho)}{4(1-\rho)^2} \qquad \blacksquare$$

(f)

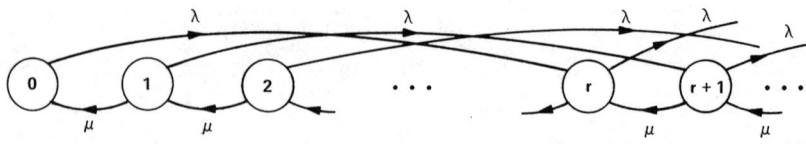

4.5.

The equilibrium equations for p_k are

$$\lambda p_0 = \mu p_1 \quad k=0$$

$$(\lambda+\mu)p_k = \mu p_{k+1} \quad 1 \leq k \leq r-1$$

$$(\lambda+\mu)p_k = \lambda p_{k-r} + \mu p_{k+1} \quad k \geq r$$

Multiply the kth equation by z^k and sum

$$\lambda \sum_{k=0}^{\infty} p_k z^k + \mu \sum_{k=1}^{\infty} p_k z^k = \lambda \sum_{k=r}^{\infty} p_{k-r} z^k + \mu \sum_{k=0}^{\infty} p_{k+1} z^k$$

$$\lambda P(z) + \mu [P(z) - p_0] = \lambda z^r P(z) + \frac{\mu}{z}[P(z) - p_0]$$

$$P(z) = \frac{\mu p_0 (z-1)}{\mu(z-1) - \lambda z(z^r - 1)}$$

$$P(z) = \frac{\mu p_0}{\mu - \lambda z \sum_{k=0}^{r-1} z^k}$$

As $\rho = \frac{r\lambda}{\mu}$ ($\bar{\lambda} = r\lambda$ and so $\rho = \bar{\lambda}\bar{x} = \frac{r\lambda}{\mu}$) we may write

$$P(z) = \frac{rp_0}{r - \rho \sum_{k=1}^{r} z^k}$$

Also $P(1) = 1 = \dfrac{rp_0}{r - r\rho} \Rightarrow p_0 = 1 - \rho.$

$$P(z) = \frac{r(1-\rho)}{r - \rho \sum_{k=1}^{r} z^k}$$

For the bulk arrival system with constant bulk size r, we have $G(z) = z^r$. Substituting this into Eq. (4.49) and simplifying gives as before

$$P(z) = \frac{r(1-\rho)}{r - \rho \sum_{k=1}^{r} z^k}$$

4.5.

To find $\overline{N}$ we note that

$$\overline{N} = \left.\frac{dP(z)}{dz}\right|_{z=1} = r(1-\rho)\rho \left.\frac{\sum_{k=1}^{r} kz^{k-1}}{\left(r-\rho\sum_{k=1}^{r} z^k\right)^2}\right|_{z=1}$$

$$= r(1-\rho)\rho \frac{r(r+1)/2}{(r-r\rho)^2}$$

$$\overline{N} = \frac{r+1}{2}\frac{\rho}{1-\rho} \qquad \blacksquare$$

To find σ_N^2 we first obtain

$$\overline{N^2} - \overline{N} = \left.\frac{d^2P(z)}{dz^2}\right|_{z=1} = r(1-\rho)\rho \left[\frac{(r-r\rho)\sum_{k=1}^{r} k(k-1) + 2\rho\left[\sum_{k=1}^{r} k\right]^2}{(r-r\rho)^3}\right]$$

Now recall that $\sum_{k=1}^{r} k = \frac{r(r+1)}{2}$ and also $\sum_{k=1}^{r} k^2 = \frac{r(r+1)(2r+1)}{6}$.

Thus $\sum_{k=1}^{r} k(k-1) = \frac{(r-1)r(r+1)}{3}$ and

$$\overline{N^2} - \overline{N} = r(1-\rho)\rho \left[\frac{r(1-\rho)\frac{(r-1)r(r+1)}{3} + 2\rho\left[\frac{r(r+1)}{2}\right]^2}{[r(1-\rho)]^3}\right]$$

$$= \frac{(r+1)\rho}{6(1-\rho)^2}(2r-2+\rho r+5\rho)$$

and so

$$\sigma_N^2 = (\overline{N^2} - \overline{N}) + \overline{N} - (\overline{N})^2$$

$$= \frac{(r+1)\rho}{6(1-\rho)^2}(2r-2+\rho r+5\rho) + \frac{(r+1)\rho}{2(1-\rho)} - \frac{(r+1)^2\rho^2}{4(1-\rho)^2}$$

$$\sigma_N^2 = \frac{(r+1)\rho}{12(1-\rho)^2}(4r+2-\rho r+\rho) \qquad \blacksquare$$

4.5.

PROBLEM 4.6.

Consider an M/M/1 queueing system with parameters λ and μ. At each of the arrival instants one new customer will enter the system with probability 1/2 or two new customers will enter simultaneously with probability 1/2.
(a) Draw the state-transition-rate diagram for this system.
(b) Using the method of non-nearest-neighbor systems write down the equilibrium equations for p_k.
(c) Find $P(z)$ and also evaluate any constants in this expression so that $P(z)$ is given in terms only of λ and μ. If possible eliminate any common factors in the numerator and denominator of this expression [this makes life simpler for you in part (d)].
(d) From part (c) find the expected number of customers in the system.
(e) Repeat part (c) using the results obtained in Section 4.5 directly.

SOLUTION

(a)

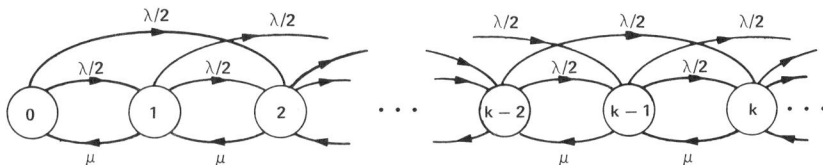

(b)
$$\lambda p_0 = \mu p_1 \quad k=0$$

$$(\lambda+\mu)p_1 = \frac{\lambda}{2}p_0 + \mu p_2 \quad k=1$$

$$(\lambda+\mu)p_k = \frac{\lambda}{2}p_{k-2} + \frac{\lambda}{2}p_{k-1} + \mu p_{k+1} \quad k \geq 2$$

(c) Multiply the kth equation by z^k and sum

$$\lambda \sum_{k=0}^{\infty} p_k z^k + \mu \sum_{k=1}^{\infty} p_k z^k = \frac{\lambda}{2} \sum_{k=2}^{\infty} p_{k-2} z^k + \frac{\lambda}{2} \sum_{k=1}^{\infty} p_{k-1} z^k + \mu \sum_{k=0}^{\infty} p_{k+1} z^k$$

$$\lambda P(z) + \mu[P(z) - p_0] = \frac{\lambda}{2} z^2 P(z) + \frac{\lambda}{2} z P(z) + \frac{\mu}{z}[P(z) - p_0]$$

$$P(z) = \frac{\mu p_0 \left(1 - \frac{1}{z}\right)}{\mu\left(1 - \frac{1}{z}\right) + \lambda\left(1 - \frac{z}{2} - \frac{z^2}{2}\right)}$$

$$P(z) = \frac{2\mu p_0}{2\mu - \lambda z(z+2)}$$

Note that $\bar{\lambda} = \frac{3}{2}\lambda$ and so $\rho = \bar{\lambda}\bar{x} = \frac{3}{2}\frac{\lambda}{\mu}$. Thus

$$P(z) = \frac{p_0}{1 - \frac{1}{3}\rho z(z+2)}$$

Also $P(1) = 1 = \frac{p_0}{1-\rho} \Rightarrow p_0 = 1 - \rho$.

$$\therefore P(z) = \frac{1-\rho}{1 - \frac{1}{3}\rho z(z+2)} \quad \blacksquare$$

(d)

$$\bar{N} = \left.\frac{dP(z)}{dz}\right|_{z=1} = (1-\rho) \left.\frac{\frac{1}{3}\rho(2z+2)}{\left[1 - \frac{1}{3}\rho z(z+2)\right]^2}\right|_{z=1}$$

$$\bar{N} = \frac{4}{3}\frac{\rho}{1-\rho} \quad \blacksquare$$

(e) Eq. (4.49) says that

$$P(z) = \frac{\mu(1-\rho)(1-z)}{\mu(1-z) - \lambda z[1 - G(z)]}$$

In our case, $g_1 = g_2 = 1/2$ and therefore $G(z) = \frac{1}{2}z + \frac{1}{2}z^2$. Thus $G^{(1)}(z) = \frac{1}{2} + z$ and $\rho = \frac{\lambda G^{(1)}(1)}{\mu} = \frac{3}{2}\frac{\lambda}{\mu}$. Substituting these values into Eq. (4.49) gives

$$P(z) = \frac{\mu(1-\rho)(1-z)}{\mu(1-z) - \lambda z\left(1 - \frac{1}{2}z - \frac{1}{2}z^2\right)}$$

4.6.

Simplifying we have

$$P(z) = \frac{1-\rho}{1 - \frac{1}{3}\rho z(z+2)}$$

as before. ■

PROBLEM 4.7.

For the bulk arrival system of Section 4.5, assume (for $0 < \alpha < 1$) that

$$g_i = (1-\alpha)\alpha^i \quad i = 0, 1, 2, \ldots$$

Find p_k = equilibrium probability of finding k in the system.

SOLUTION

Since $g_i = (1-\alpha)\alpha^i$, $i = 0, 1, 2, \ldots$ the appropriate z-transform for the distribution of bulk size is (see footnote on page 135)

$$G(z) = \sum_{i=0}^{\infty} g_i z^i = \sum_{i=0}^{\infty} (1-\alpha)\alpha^i z^i.$$

$$G(z) = \frac{1-\alpha}{1-\alpha z} \quad ■$$

Further, we have $\rho = \dfrac{\lambda G^{(1)}(1)}{\mu} = \dfrac{\lambda \alpha}{\mu(1-\alpha)}$. Therefore, Eq. (4.49) gives

$$P(z) = \frac{\mu(1-\rho)(1-z)}{\mu(1-z) - \lambda z[1-G(z)]}$$

$$= \frac{\mu(1-\rho)(1-z)}{\mu(1-z) - \lambda z\left[1 - \dfrac{1-\alpha}{1-\alpha z}\right]}$$

$$= \frac{\mu(1-\rho)(1-\alpha z)}{\mu(1-\alpha z) - \lambda \alpha z}$$

$$P(z) = (1-\rho) \frac{1-\alpha z}{1 - \dfrac{\mu+\lambda}{\mu}\alpha z} \quad ■$$

To invert $P(z)$, we note that the numerator degree equals the denominator degree, and so we divide once to bring the expression into proper form as follows:

$$P(z) = (1-\rho)\left[\frac{\mu}{\lambda+\mu} + \frac{\lambda}{\lambda+\mu}\frac{1}{1-\frac{\mu+\lambda}{\mu}\alpha z}\right]$$

By inspection, we have the final answer

$$p_0 = 1-\rho$$

$$p_k = (1-\rho)\frac{\lambda}{\lambda+\mu}\left(\frac{\mu+\lambda}{\mu}\alpha\right)^k \quad k \geq 1$$

∎

PROBLEM 4.8.

For the bulk arrival system studied in Section 4.5, find the mean $\overline{N}$ and variance σ_N^2 for the number of customers in the system. Express your answers in terms of the moments of the bulk arrival distribution.

SOLUTION

Eq. (4.49) gives

$$P(z) = \frac{\mu(1-\rho)(1-z)}{\mu(1-z) - \lambda z[1-G(z)]} = \frac{N(z)}{D(z)}$$

Let the kth moment of the bulk size be denoted by $\overline{g^k}$. Note that $\overline{g} = G^{(1)}(1)$ and $\overline{g^2} - \overline{g} = G^{(2)}(1)$. From Eq. (4.50) we have $\rho = \lambda \overline{g}/\mu$. To find $\overline{N}$ we use

$$\overline{N} = \frac{dP(z)}{dz}\bigg|_{z=1} = \frac{D(z) N^{(1)}(z) - N(z) D^{(1)}(z)}{[D(z)]^2}\bigg|_{z=1}$$

This last is indeterminate (since $N(1) = D(1) = 0$), and so we must use L'Hospital's rule twice as follows:

$$\overline{N} = P^{(1)}(1) = \frac{D(z) N^{(2)}(z) - N(z) D^{(2)}(z)}{2 D^{(1)}(z) D(z)}\bigg|_{z=1}$$

$$= \frac{D(z) N^{(3)}(z) + D^{(1)}(z) N^{(2)}(z) - N^{(1)}(z) D^{(2)}(z) - N(z) D^{(3)}(z)}{2 D^{(2)}(z) D(z) + 2[D^{(1)}(z)]^2}\bigg|_{z=1}$$

Now since

4.7. – 4.8.

$$N(z) = \mu(1-\rho)(1-z) \qquad N(1) = 0$$
$$N^{(1)}(z) = -\mu(1-\rho) \qquad N^{(1)}(1) = -\mu(1-\rho)$$
$$N^{(2)}(z) = 0 \qquad N^{(2)}(1) = 0$$
$$N^{(3)}(z) = 0 \qquad N^{(3)}(1) = 0$$

and also

$$D(z) = \mu(1-z) - \lambda z[1 - G(z)] \qquad D(1) = 0$$
$$D^{(1)}(z) = -\mu - \lambda + \lambda G(z) + \lambda z G^{(1)}(z) \qquad D^{(1)}(1) = -\mu(1-\rho)$$
$$D^{(2)}(z) = 2\lambda G^{(1)}(z) + \lambda z G^{(2)}(z) \qquad D^{(2)}(1) = \lambda(\overline{g^2} + \overline{g})$$
$$D^{(3)}(z) = 3\lambda G^{(2)}(z) + \lambda z G^{(3)}(z) \qquad D^{(3)}(1) = \lambda(\overline{g^3} - \overline{g})$$

we may substitute in our expression for $\overline{N}$ to obtain

$$\overline{N} = \frac{\mu(1-\rho)\lambda(\overline{g^2}-\overline{g})}{2[-\mu(1-\rho)]^2} = \frac{\lambda(\overline{g^2}-\overline{g})}{2\mu(1-\rho)}$$

$$\overline{N} = \frac{\rho}{1-\rho}\left[\frac{\overline{g^2}+\overline{g}}{2\overline{g}}\right] \qquad \blacksquare$$

For the variance, we use

$$\sigma_N^2 = \left.\frac{d^2 P(z)}{dz^2}\right|_{z=1} + \overline{N} - (\overline{N})^2$$

We see that

$$P^{(2)}(1) = \left.\frac{D(z)[D(z)N^{(2)}(z) - N(z)D^{(2)}(z)]}{[D(z)]^3}\right|_{z=1}$$

$$- \left.\frac{2D^{(1)}(z)[D(z)N^{(1)}(z) - N(z)D^{(1)}(z)]}{[D(z)]^3}\right|_{z=1}$$

We must now apply L'Hospital's rule three times, evaluate at $z=1$, and eliminate the terms that vanish. We obtain

$$P^{(2)}(1) = \frac{-N^{(1)}(1)[2D^{(1)}(1)D^{(3)}(1) - 3[D^{(2)}(1)]^2]}{6[D^{(1)}(1)]^3}$$

$$= \frac{\mu(1-\rho)\left[-2\mu(1-\rho)\lambda(\overline{g^3}-\overline{g}) - 3\lambda^2(\overline{g^2}+\overline{g})^2\right]}{-6[\mu(1-\rho)]^3}$$

$$= \frac{\lambda(\overline{g^3}-\overline{g})}{3\mu(1-\rho)} + \frac{\lambda^2(\overline{g^2}+\overline{g})^2}{2\mu^2(1-\rho)^2}$$

4.8.

Hence

$$\sigma_N{}^2 = \frac{\lambda(\overline{g^3}-\overline{g})}{3\mu(1-\rho)} + \frac{\lambda^2(\overline{g^2}+\overline{g})^2}{2\mu^2(1-\rho)^2} + \frac{\lambda(\overline{g^2}+\overline{g})}{2\mu(1-\rho)} - \frac{\lambda^2(\overline{g^2}+\overline{g})^2}{4\mu^2(1-\rho)^2}$$

Thus

$$\sigma_N{}^2 = \frac{\rho}{1-\rho}\left[\frac{2\overline{g^3}+3\overline{g^2}+\overline{g}}{6\overline{g}}\right] + \frac{\rho^2}{(1-\rho)^2}\left[\frac{\overline{g^2}+\overline{g}}{2\overline{g}}\right]^2 \qquad \blacksquare$$

or

$$\sigma_N{}^2 = \frac{\rho}{1-\rho}\left[\frac{2\overline{g^3}+3\overline{g^2}+\overline{g}}{6\overline{g}}\right] + (\overline{N})^2 \qquad \blacksquare$$

PROBLEM 4.9.

Consider an M/M/1 system with the following variation: Whenever the server becomes free, he accepts *two* customers (if at least two are available) from the queue into service simultaneously. Of these two customers, only one receives service; when the service for this one is completed, both customers depart (and so the other customer got a "free ride").

If only one customer is available in the queue when the server becomes free, then that customer is accepted alone and is serviced; if a new customer happens to arrive when this single customer is being served, then the new customer joins the old one in service and this new customer receives a "free ride".

In all cases, the service time is exponentially distributed with mean $1/\mu$ sec and the average (Poisson) arrival rate is λ customers per second.

(a) Draw the appropriate state diagram.
(b) Write down the appropriate difference equations for p_k = equilibrium probability of finding k customers in the system.
(c) Solve for $P(z)$ in terms of p_0 and p_1.
(d) Express p_1 in terms of p_0.

SOLUTION

(a)

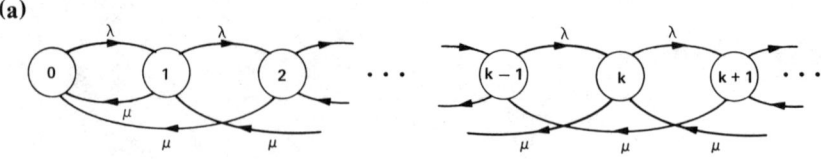

(b)
$$\lambda p_0 = \mu p_1 + \mu p_2 \quad k=0$$
$$(\lambda+\mu)p_k = \lambda p_{k-1} + \mu p_{k+2} \quad k \geq 1$$

(c) Multiply the kth equation by z^k and sum

$$\lambda \sum_{k=0}^{\infty} p_k z^k + \mu \sum_{k=1}^{\infty} p_k z^k = \lambda \sum_{k=1}^{\infty} p_{k-1} z^k + \mu \sum_{k=0}^{\infty} p_{k+2} z^k + \mu p_1$$

$$\lambda P(z) + \mu[P(z) - p_0] = \lambda z P(z) + \frac{\mu}{z^2}[P(z) - p_0 - p_1 z] + \mu p_1$$

$$P(z) = \frac{\mu p_0(z+1) + \mu p_1 z}{\mu(z+1) - \lambda z^2}$$

(d) Since $P(1) = 1 = \dfrac{2\mu p_0 + \mu p_1}{2\mu - \lambda}$ we have

$$p_1 = 2(1 - p_0) - \frac{\lambda}{\mu}$$

Thus

$$P(z) = \frac{(2\mu - \lambda)z + \mu p_0(1-z)}{\mu(z+1) - \lambda z^2}$$

[Note: Proceeding beyond that required by the problem statement, we now solve for p_k explicitly. The two roots of the denominator of $P(z)$ are

$$z_1 = \frac{\mu - \sqrt{\mu^2 + 4\lambda\mu}}{2\lambda}, \quad z_2 = \frac{\mu + \sqrt{\mu^2 + 4\lambda\mu}}{2\lambda}$$

For $\lambda > 0$, we have

$$|z_1| = -z_1 = \frac{\sqrt{\mu^2 + 4\lambda\mu} - \mu}{2\lambda} < \frac{\sqrt{\mu^2 + 4\lambda\mu + 4\lambda^2} - \mu}{2\lambda} = 1$$

Thus the analyticity of $P(z)$ for $|z| \leq 1$ requires that the one root of the numerator of $P(z)$ must be the denominator root z_1 (as $|z_1| < 1$), and so an equilibrium solution exists iff the other denominator root z_2 satisfies $|z_2| > 1$. We see that $|z_2| > 1$ if $\lambda < 2\mu$ as follows:

$$|z_2| = z_2 = \frac{\mu + \sqrt{\mu^2 + 4\lambda\mu}}{2\lambda} > \frac{(\lambda/2) + \sqrt{(\lambda^2/4) + 2\lambda^2}}{2\lambda} = 1$$

On the other hand, if $\lambda \geq 2\mu$, then $|z_2| \leq 1$; hence we observe that the system is stable iff $\lambda < 2\mu$. Now let us find p_k. For $\lambda < 2\mu$, z_1 is the root of the numerator and so

$$(2\mu - \lambda) z_1 + \mu p_0 (1 - z_1) = 0$$

or

$$p_0 = \frac{(2\mu - \lambda) z_1}{\mu (z_1 - 1)}$$

Substituting this into $P(z)$ (and canceling the root at $z = z_1$) we obtain

$$P(z) = \frac{2\mu - \lambda}{\lambda (1 - z_1) z_2} \frac{1}{1 - (z/z_2)}$$

But the product of the roots is simply $z_1 z_2 = -\frac{\mu}{\lambda}$; using this and the fact that $|z_2| > 1$ we invert by inspection to obtain

$$p_k = \frac{2\mu - \lambda}{\mu + \lambda z_2} \left(\frac{1}{z_2} \right)^k \quad k \geq 0$$

as the desired solution.]

PROBLEM 4.10.†

We consider the denominator polynomial in Eq. (4.35).
(a) Of the $r + 1$ roots, one occurs at $z = 1$. Use Rouche's theorem to show that exactly r roots lie in the unit disk $|z| \leq 1$.
(b) Show that $z = 1$ is the only root on the unit circle $|z| = 1$.
[Hence (a) and (b) establish that exactly $r - 1$ roots lie in the range $|z| < 1$, and one root, say z_0, lies in the region $|z_0| > 1$.]

SOLUTION

(a) Eq. (4.35) is

$$P(z) = \frac{(1 - z^r) \sum_{j=0}^{r-1} P_j z^j}{r\rho z^{r+1} - (1 + r\rho) z^r + 1} = \frac{N(z)}{D(z)}$$

We split the denominator into two parts, $D(z) = f(z) + g(z)$, where

$$f(z) = -(1 + r\rho) z^r, \quad g(z) = r\rho z^{r+1} + 1.$$

For $\rho < 1$, we choose any real δ such that $0 < \delta < \frac{1-\rho}{r\rho}$. (Note that $0 < 1 - \rho - r\rho \delta$.) Define the closed contour C as a circle about the origin of radius $1 + \delta$. On C, $|z| = 1 + \delta$, so we have

$$|f(z)| = |-(1+r\rho)z^r| = (1+r\rho)(1+\delta)^r$$

and

$$|g(z)| = |r\rho z^{r+1}+1| \leq r\rho(1+\delta)^{r+1}+1$$

Thus, on the contour C, we have

$$|f(z)|-|g(z)| \geq (1+r\rho)(1+\delta)^r - r\rho(1+\delta)^{r+1} - 1$$

$$= (1+\delta)^r(1-r\rho\delta) - 1$$

$$\geq (1+r\delta)(1-r\rho\delta) - 1$$

(since $(1+\delta)^r \geq 1+r\delta$)

$$= r\delta(1-\rho-r\rho\delta) > 0$$

by choice of δ. So $|f(z)| > |g(z)|$ on C, and by Rouche's theorem, the denominator polynomial $f(z)+g(z)$ has exactly r roots in the range $|z| < 1+\delta$ since $f(z)$ has. As this is true for all $0 < \delta < \frac{1-\rho}{r\rho}$, letting $\delta \to 0$, we see that the denominator polynomial has exactly r roots in the range $|z| \leq 1$.

[Note that the only root of the denominator which is also a root of $1-z^r$ is, by substitution, $z=1$. Thus the $r-1$ remaining roots must cancel with the $r-1$ roots of $\sum_{j=0}^{r-1} P_j z^j$ as stated in the text. However, it was also mentioned there that these $r-1$ roots all satisfy $|z| < 1$. This is easily established below in part (b), although it is not really needed for the argument in the text to go through.]

(b) Assume $|z|=1$ and $r\rho z^{r+1} - (1+r\rho)z^r + 1 = 0$. So $(1+r\rho - r\rho z)z^r = 1$ and thus $|1+r\rho(1-z)| = 1$. Define $h = 1+r\rho(1-z)$. Since we have $-1 \leq \text{Re}(z) \leq 1$, then $\text{Re}(h) = 1+r\rho[1-\text{Re}(z)] \geq 1$. Thus

$$1 = |h| = \sqrt{[\text{Re}(h)]^2+[\text{Im}(h)]^2} \geq \text{Re}(h) \geq 1$$

and so

$$1 = \text{Re}(h) = 1+r\rho[1-\text{Re}(z)].$$

$$\therefore \text{Re}(z) = 1$$

As $|z| = 1$ and $\text{Re}(z) = 1$, we have $\text{Im}(z) = 0$.

$$\therefore z = 1$$

4.10.

PROBLEM 4.11.

Show that the solution to Eq. (4.71) gives a set of variables $\{x_i\}$ which guarantee that Eq. (4.72) is indeed the solution to Eq. (4.69).

SOLUTION

We proceed by showing that for Eq. (4.69)' the left-hand side (LHS) equals the right-hand side (RHS) under the proposed solution Eq. (4.72) if the $\{x_i\}$ satisfy Eq. (4.71). Noting that $\delta_{k_i-1}\alpha_i(k_i) = \alpha_i(k_i)$ for each i, the LHS can be written as

$$\text{LHS} = p(k_1, \ldots, k_N)\sum_{i=1}^{N}\alpha_i(k_i)\mu_i.$$

The RHS is

$$\sum_{i=1}^{N}\sum_{j=1}^{N}\delta_{k_j-1}\alpha_i(k_i+1)\mu_i r_{ij} p(k_1, \ldots, k_j-1, \ldots, k_i+1, \ldots, k_N).$$

Using the proposed solution

$$p(k_1, \ldots, k_N) = \frac{1}{G(K)}\prod_{l=1}^{N}\frac{x_l^{k_l}}{\beta_l(k_l)}$$

we may write

$$p(k_1, \ldots, k_j-1, \ldots, k_i+1, \ldots, k_N) = \frac{1}{G(K)}\prod_{l=1}^{N}\frac{x_l^{k_l}}{\beta_l(k_l)}$$

$$\cdot \frac{\dfrac{x_i}{x_j}}{\dfrac{\beta_i(k_i+1)}{\beta_i(k_i)} \cdot \dfrac{\beta_j(k_j-1)}{\beta_j(k_j)}}$$

$$= p(k_1, \ldots, k_N) \frac{x_i}{x_j} \cdot \frac{\beta_i(k_i)}{\beta_i(k_i+1)} \cdot \frac{\beta_j(k_j)}{\beta_j(k_j-1)}$$

Since $\beta_i(k_i+1) = \beta_i(k_i)\alpha_i(k_i+1)$ and $\beta_j(k_j) = \beta_j(k_j-1)\alpha_j(k_j)$ we have

$$p(k_1, \ldots, k_j-1, \ldots, k_i+1, \ldots, k_N) = p(k_1, \ldots, k_N)\frac{x_i}{x_j} \cdot \frac{\alpha_j(k_j)}{\alpha_i(k_i+1)}$$

4.11.

Thus the RHS may be expressed as

$$\text{RHS} = \sum_{i=1}^{N}\sum_{j=1}^{N} \delta_{k_j-1}\alpha_i(k_i+1)\mu_i r_{ij} p(k_1, \ldots, k_N) \frac{x_i}{x_j} \cdot \frac{\alpha_j(k_j)}{\alpha_i(k_i+1)}$$

$$= p(k_1, \ldots, k_N) \sum_{i=1}^{N}\sum_{j=1}^{N} \delta_{k_j-1}\alpha_j(k_j)\mu_i r_{ij} \frac{x_i}{x_j}$$

But since $\delta_{k_j-1}\alpha_j(k_j) = \alpha_j(k_j)$,

$$\text{RHS} = p(k_1, \ldots, k_N) \sum_{i=1}^{N}\sum_{j=1}^{N} \alpha_j(k_j)\mu_i r_{ij} \frac{x_i}{x_j}$$

Thus the LHS equals the RHS if

$$\sum_{i=1}^{N} \alpha_i(k_i)\mu_i = \sum_{i=1}^{N}\sum_{j=1}^{N} \alpha_j(k_j)\mu_i r_{ij} \frac{x_i}{x_j}.$$

We prove this last equality by applying Eq. (4.71). That is,

$$\mu_i = \sum_{j=1}^{N} \mu_j r_{ji} \frac{x_j}{x_i} \quad i=1,2,\ldots,N.$$

Multiplying each of these N equations by $\alpha_i(k_i)$, summing on i, and then interchanging the indices i and j gives

$$\sum_{i=1}^{N} \alpha_i(k_i)\mu_i = \sum_{i=1}^{N}\sum_{j=1}^{N} \alpha_j(k_j)\mu_i r_{ij} \frac{x_i}{x_j}$$

which was to be shown. This completes the proof.

PROBLEM 4.12.

(a) Draw the state-transition-rate diagram showing local balance for the case $(N = 3, K = 5)$ with the following structure:

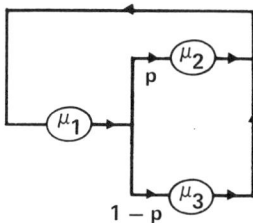

(b) Solve for $p(k_1, k_2, k_3)$.

SOLUTION

The state of the system is represented by the circled triplet

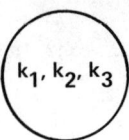

where k_i = the number in the ith node ($i = 1, 2, 3$) and where the k_i must satisfy $k_1 + k_2 + k_3 = K = 5$. Let $q = 1 - p$. Recall that a local balance equation (with respect to a given network state and a network node i) equates the rate of flow out of that network state due to the departure of a customer from node i to the rate of flow into that network state due to the arrival of a customer to node i. In the state-transition-rate diagram below, a set of flows to be balanced in a local balance equation is joined by a heavy black line.
(a)

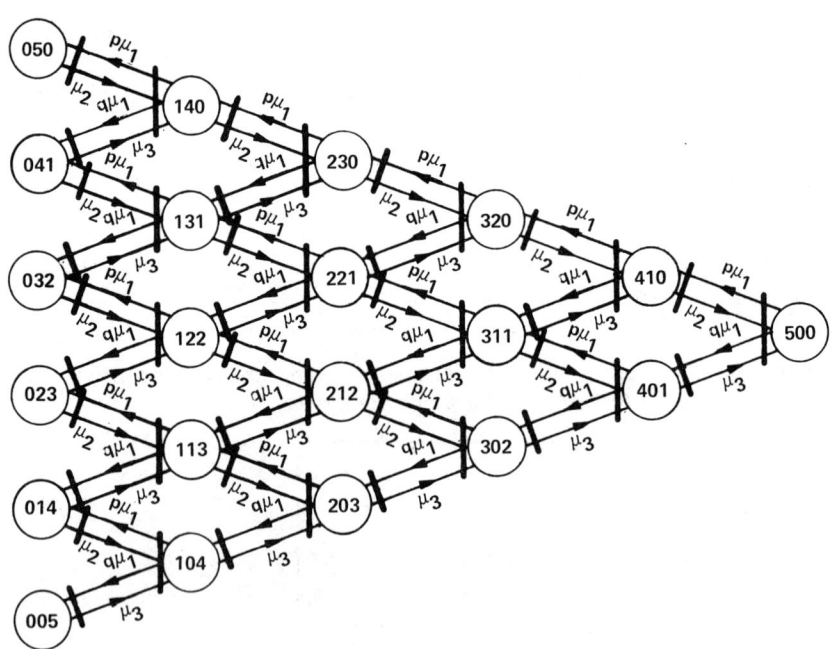

4.12.

(b) The local balance equations among states for which $k_3=0$ are:
$$\mu_2 p(0\,5\,0) = \mu_1 p\, p(1\,4\,0)$$
$$\mu_2 p(1\,4\,0) = \mu_1 p\, p(2\,3\,0)$$
$$\mu_2 p(2\,3\,0) = \mu_1 p\, p(3\,2\,0)$$
$$\mu_2 p(3\,2\,0) = \mu_1 p\, p(4\,1\,0)$$
$$\mu_2 p(4\,1\,0) = \mu_1 p\, p(5\,0\,0)$$

which give
$$p(5-k, k, 0) = \left(\frac{\mu_1 p}{\mu_2}\right)^k p(5\,0\,0) \quad \text{for} \quad k=1,2,3,4,5$$

The equations when $k_2=0$ are:
$$\mu_3 p(0\,0\,5) = \mu_1 q\, p(1\,0\,4)$$
$$\mu_3 p(1\,0\,4) = \mu_1 q\, p(2\,0\,3)$$
$$\mu_3 p(2\,0\,3) = \mu_1 q\, p(3\,0\,2)$$
$$\mu_3 p(3\,0\,2) = \mu_1 q\, p(4\,0\,1)$$
$$\mu_3 p(4\,0\,1) = \mu_1 q\, p(5\,0\,0)$$

which give
$$p(5-k, 0, k) = \left(\frac{\mu_1 q}{\mu_3}\right)^k p(5\,0\,0) \quad \text{for} \quad k=1,2,3,4,5$$

Similarly, when $k_3=1$ we obtain
$$p(4-k, k, 1) = \left(\frac{\mu_1 p}{\mu_2}\right)^k \left(\frac{\mu_1 q}{\mu_3}\right) p(5\,0\,0) \quad \text{for} \quad k=1,2,3,4$$

and when $k_2=1$ we find
$$p(4-k, 1, k) = \left(\frac{\mu_1 p}{\mu_2}\right)\left(\frac{\mu_1 q}{\mu_3}\right)^k p(5\,0\,0) \quad \text{for} \quad k=1,2,3,4$$

Finally, we get
$$p(k_1, k_2, k_3) = \left(\frac{\mu_1 p}{\mu_2}\right)^{k_2} \left(\frac{\mu_1 q}{\mu_3}\right)^{k_3} p(5\,0\,0).$$

To find $p(5\,0\,0)$, we use $\sum_{k_1+k_2+k_3=5} p(k_1,k_2,k_3) = 1$. We may eliminate k_1 by observing that for any $0 \leqslant k_2 \leqslant 5$, $0 \leqslant k_3 \leqslant 5-k_2$ we must have $k_1 = 5 - k_3 - k_2$. Thus

4.12.

$$\sum_{k_1+k_2+k_3=5} p(k_1, k_2, k_3) = \sum_{k_2=0}^{5} \sum_{k_3=0}^{5-k_2} p(5-k_2-k_3, k_2, k_3)$$

and so

$$p(5\,0\,0) = \frac{1}{\sum_{k_2=0}^{5} \sum_{k_3=0}^{5-k_2} \left(\frac{\mu_1 p}{\mu_2}\right)^{k_2} \left(\frac{\mu_1 q}{\mu_3}\right)^{k_3}}$$

Hence

$$p(k_1, k_2, k_3) = \frac{\left(\frac{\mu_1 p}{\mu_2}\right)^{k_2} \left(\frac{\mu_1 q}{\mu_3}\right)^{k_3}}{\sum_{k_2=0}^{5} \sum_{k_3=0}^{5-k_2} \left(\frac{\mu_1 p}{\mu_2}\right)^{k_2} \left(\frac{\mu_1 q}{\mu_3}\right)^{k_3}} \quad \blacksquare$$

where $k_1 + k_2 + k_3 = 5$.

PROBLEM 4.13.

Consider a two-node Markovian queueing network (of the more general type considered by Jackson) for which $N = 2$, $m_1 = m_2 = 1$, $\mu_{k_i} = \mu_i$ (constant service rate), and which has transition probabilities (r_{ij}) as described in the following matrix:

$$r_{ij} = \begin{array}{c|cccc} {}_i\backslash{}^j & 0 & 1 & 2 & 3 \\ \hline 0 & 0 & 1 & 0 & 0 \\ 1 & 0 & 0 & 1-\alpha & \alpha \\ 2 & 0 & 1 & 0 & 0 \end{array}$$

where $0 < \alpha < 1$ and nodes 0 and $N+1$ are the "source" and "sink" nodes, respectively. We also have (for some integer K)

$$\gamma(S(k_1, k_2)) = \begin{cases} \infty & k_1 + k_2 \neq K \\ 0 & k_1 + k_2 = K \end{cases}$$

and assume the system initially contains K customers.
(a) Find e_i ($i = 1, 2$) as given in Eq. (4.75).
(b) Since $N = 2$, let us denote $p(k_1, k_2) = p(k_1, K-k_1)$ by p_{k_1}. Find the balance equations for p_{k_1}.

4.12.–4.13.

(c) Solve these equations for p_{k_1} explicitly.
(d) By considering the fraction of time the first node is busy, find the time between customer departures from the network (via node 1, of course).

SOLUTION

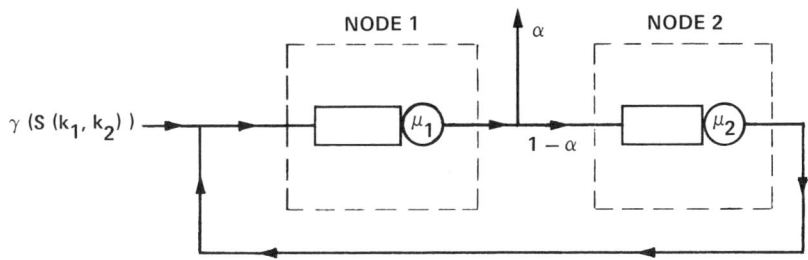

(a) Using Eq. (4.75) e_i is found as follows:

$$e_1 = r_{01} + \sum_{j=1}^{2} e_j r_{j1} \Rightarrow e_1 = 1 + e_1 \cdot 0 + e_2 \cdot 1 = 1 + e_2$$

$$e_2 = r_{02} + \sum_{j=1}^{2} e_j r_{j2} \Rightarrow e_2 = 0 + e_1 \cdot (1-\alpha) + e_2 \cdot 0 = e_1(1-\alpha)$$

Solving gives

$$e_1 = \frac{1}{\alpha}, \quad e_2 = \frac{1-\alpha}{\alpha} \qquad \blacksquare$$

(b) Since $\gamma(S(k_1, k_2)) = 0$ for $k_1 + k_2 = K$, no-one enters the system if K customers are already there. But as soon as a departure takes place, another customer immediately enters the system (since $\gamma(S(k_1, k_2)) = \infty$ for $k_1 + k_2 \neq K$). Thus we have a *closed* queueing network as follows:

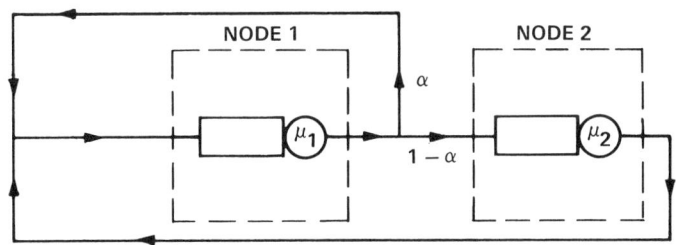

4.13.

For this network we have the following state diagram (labeling the states only by the number k_1 present at node 1):

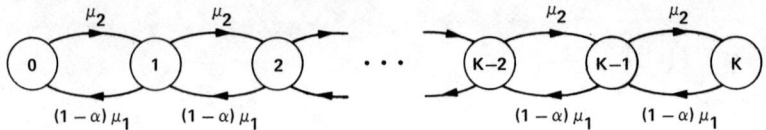

The balance equations for p_{k_1} are:

$$\mu_2 p_0 = (1-\alpha)\mu_1 p_1 \quad k_1 = 0 \qquad \blacksquare$$

$$[\mu_2 + (1-\alpha)\mu_1] p_{k_1} = \mu_2 p_{k_1-1} + (1-\alpha)\mu_1 p_{k_1+1} \quad 0 < k_1 < K \qquad \blacksquare$$

$$(1-\alpha)\mu_1 p_K = \mu_2 p_{K-1} \quad k_1 = K \qquad \blacksquare$$

(c) Noting that this is the same state diagram as Fig. 3.8 for M/M/1/K (and also for Exercise 3.12), we immediately solve for p_{k_1} from Eq. (3.43).

$$p_{k_1} = \frac{1 - \dfrac{\mu_2}{(1-\alpha)\mu_1}}{1 - \left[\dfrac{\mu_2}{(1-\alpha)\mu_1}\right]^{K+1}} \left[\dfrac{\mu_2}{(1-\alpha)\mu_1}\right]^{k_1} \quad 0 \leq k_1 \leq K \qquad \blacksquare$$

(d) The first node is busy a fraction $(1 - p_0)$ of the time. For a very long time interval τ, the first node is busy for $(1 - p_0)\tau$ seconds. While node 1 is busy, customers leave the system at rate $\alpha\mu_1$. Thus $\alpha\mu_1(1 - p_0)\tau$ is the average number of departures during τ. Therefore, by renewal theory arguments, the average time between departures will be $\dfrac{1}{\alpha\mu_1(1-p_0)}$ or, upon substituting for p_0, the average interdeparture time is

$$\frac{1-\alpha}{\alpha\mu_2} \frac{1 - \left[\dfrac{\mu_2}{(1-\alpha)\mu_1}\right]^{K+1}}{1 - \left[\dfrac{\mu_2}{(1-\alpha)\mu_1}\right]^K} \qquad \blacksquare$$

4.13.

Chapter 5

The Queue M/G/1

PROBLEM 5.1.

Prove Eq. (5.14) from Eq. (5.11).

SOLUTION

We first derive a power series expansion for the Laplace transform of any nonnegative random variable in terms of its moments. To this end, let $X \geq 0$ be a random variable with density $h(x)$, Laplace transform $H^*(s)$, and kth moment $\overline{X^k}$. Recalling a technique from Appendix II on page 382 and expanding the exponential in a power series, the Laplace transform is given by

$$H^*(s) \triangleq \int_0^\infty e^{-sx} h(x)\, dx = \int_0^\infty \sum_{k=0}^\infty \frac{(-sx)^k}{k!} h(x)\, dx$$

$$= \sum_{k=0}^\infty \frac{(-1)^k s^k}{k!} \int_0^\infty x^k h(x)\, dx = \sum_{k=0}^\infty \frac{(-1)^k \overline{X^k}}{k!} s^k$$

or

$$H^*(s) = 1 + \sum_{k=1}^\infty \frac{(-1)^k \overline{X^k}}{k!} s^k$$

[Note that this immediately gives Eq. (II.26), namely, $H^{*(n)}(0) = (-1)^n \overline{X^n}$.] Applying the above expansion to $\hat{F}^*(s)$, the transform of the residual life density, we find

$$\hat{F}^*(s) = 1 + \sum_{k=1}^\infty \frac{(-1)^k r_k}{k!} s^k$$

But Eq. (5.11) gives

$$\hat{F}^*(s) = \frac{1 - F^*(s)}{s\, m_1}.$$

5.1.

Using the expansion for $F^*(s)$ in this equation we have

$$\hat{F}^*(s) = \frac{1 - \left[1 + \sum_{k=1}^{\infty} \frac{(-1)^k m_k}{k!} s^k\right]}{s \, m_1} = \sum_{k=0}^{\infty} \frac{(-1)^k m_{k+1}}{(k+1)! \, m_1} s^k$$

$$\hat{F}^*(s) = 1 + \sum_{k=1}^{\infty} \frac{(-1)^k m_{k+1}}{(k+1)! \, m_1} s^k$$

Equating coefficients of these two power series expansions for $\hat{F}^*(s)$ gives

$$r_k = \frac{m_{k+1}}{(k+1) \, m_1} \qquad k = 1, 2, 3, \ldots$$

which is Eq. (5.14).

PROBLEM 5.2†

Here we derive the residual lifetime density $\hat{f}(x)$ discussed in Section 5.2. We use the notation of Figure 5.1.

(a) Observing that the event $\{Y \leq y\}$ can occur if and only if $t < \tau_k \leq t+y < \tau_{k+1}$ for some k, show that

$$\hat{F}_t(y) \triangleq P[Y \leq y]$$

$$= \sum_{k=1}^{\infty} \int_t^{t+y} [1 - F(t+y-x)] \, dP[\tau_k \leq x]$$

(b) Observing that $\tau_k \leq x$ if and only if $\alpha(x)$, the number of "arrivals" in $(0, x)$, is at least k, that is, $P[\tau_k \leq x] = P[\alpha(x) \geq k]$, show that

$$\sum_{k=1}^{\infty} P[\tau_k \leq x] = \sum_{k=1}^{\infty} k P[\alpha(x) = k]$$

(c) Let $\hat{F}(y) = \lim \hat{F}_t(y)$ as $t \to \infty$ with corresponding pdf $\hat{f}(y)$. Show that we now have

$$\hat{f}(y) = \frac{1 - F(y)}{m_1}$$

[HINT: Use the key renewal theorem (see [FELL 66] or [TAKA 62a], for example).]

SOLUTION

(a) The event $\{Y \leq y\}$ may be written as the following disjoint union of events

$$\{Y \leq y\} = \bigcup_{k=1}^{\infty} \{t < \tau_k \leq t+y < \tau_{k+1}\}$$

That is, the instant $t+y$ must be separated from t by *at least* one arrival; we let τ_k denote the latest of these arrival instants (therefore $t < \cdots < \tau_{k-1} < \tau_k \leq t+y < \tau_{k+1}$).

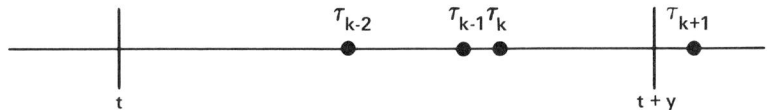

Rewriting this union we see that

$$\{Y \leq y\} = \bigcup_{k=1}^{\infty} \{t < \tau_k \leq t+y, \, t+y-\tau_k < \tau_{k+1}-\tau_k\}$$

$$= \bigcup_{k=1}^{\infty} \{t < \tau_k \leq t+y, \, t+y-\tau_k < t_{k+1}\}$$

where we use the definition $t_{k+1} \triangleq \tau_{k+1}-\tau_k$ which is the length of the $(k+1)$th interarrival interval. Thus

$$\hat{F}_t(y) = P[Y \leq y] = \sum_{k=1}^{\infty} P[t < \tau_k \leq t+y, \, t_{k+1} > t+y-\tau_k]$$

Conditioning on τ_k we obtain

$$\hat{F}_t(y) = \sum_{k=1}^{\infty} \int_0^{\infty} P[t < \tau_k \leq t+y, \, t_{k+1} > t+y-\tau_k \mid \tau_k = x]$$

$$\cdot P[x < \tau_k \leq x+dx]$$

$$= \sum_{k=1}^{\infty} \int_0^{\infty} P[t < x \leq t+y, \, t_{k+1} > t+y-x] \, dP[\tau_k \leq x]$$

$$= \sum_{k=1}^{\infty} \int_t^{t+y} P[t_{k+1} > t+y-x] \, dP[\tau_k \leq x]$$

5.2

Thus

$$\hat{F}_t(y) = \sum_{k=1}^{\infty} \int_t^{t+y} [1 - F(t+y-x)] \, dP[\tau_k \leqslant x]$$

[Note: A similar formula for $\hat{F}_t(y)$ may be obtained as follows. Recognizing that the event $\{Y \leqslant y\}$ may be written as

$$\{Y \leqslant y\} = \bigcup_{k=0}^{\infty} \{\tau_k < t,\ t \leqslant \tau_{k+1} < t+y\}$$

(where $\tau_0 \triangleq 0$), we then have

$$\hat{F}_t(y) = \sum_{k=0}^{\infty} P[\tau_k < t,\ t - \tau_k \leqslant t_{k+1} < t+y - \tau_k]$$

$$= \sum_{k=0}^{\infty} \int_0^{\infty} P[x < t,\ t - x \leqslant t_{k+1} < t+y-x] \, dP[\tau_k \leqslant x]$$

$$= \sum_{k=0}^{\infty} \int_0^{t} P[t-x \leqslant t_{k+1} < t+y-x] \, dP[\tau_k \leqslant x]$$

or

$$\hat{F}_t(y) = F(t+y) - F(t) + \sum_{k=1}^{\infty} \int_0^{t} [F(t+y-x) - F(t-x)] \, dP[\tau_k \leqslant x]$$

It can be shown that this expression for $\hat{F}_t(y)$ is equal to that derived immediately above.]

(b) Recall that for a discrete random variable $X \geqslant 0$, we have

$$E[X] = \sum_{j=1}^{\infty} j\, P[X=j] = \sum_{j=1}^{\infty} \sum_{k=1}^{j} P[X=j]$$

$$= \sum_{k=1}^{\infty} \sum_{j=k}^{\infty} P[X=j] = \sum_{k=1}^{\infty} P[X \geqslant k]$$

We are given $P[\tau_k \leqslant x] = P[\alpha(x) \geqslant k]$. Summing on k and using our result above for $E[X]$, we have

$$\sum_{k=1}^{\infty} P[\tau_k \leqslant x] = \sum_{k=1}^{\infty} P[\alpha(x) \geqslant k]$$

$$= \sum_{k=1}^{\infty} k\, P[\alpha(x) = k] = E[\alpha(x)].$$

(c) We can note (using the strong law of large numbers) that as $x \to \infty$ then $\dfrac{\alpha(x)}{x} \to \dfrac{1}{m_1}$ with probability 1. We also have $\dfrac{E[\alpha(x)]}{x} \to \dfrac{1}{m_1}$ as $x \to \infty$,

5.2

which is simply the Elementary Renewal Theorem. Thus this gives the intuitive result

$$\sum_{k=1}^{\infty} P[\tau_k \leqslant x] = E[\alpha(x)] \approx \frac{x}{m_1} \quad \text{for large } x.$$

To give a rigorous proof of part (c), we proceed as follows. We first simplify $\hat{F}_t(y)$ as given in part (a):

$$\hat{F}_t(y) = \int_t^{t+y} [1 - F(t+y-x)] \, d \sum_{k=1}^{\infty} P[\tau_k \leqslant x]$$

$$= \int_t^{t+y} [1 - F(t+y-x)] \, d E[\alpha(x)] \quad \text{(from part (b))}$$

$$= \int_0^{t+y} [1 - F(t+y-x)] \, d E[\alpha(x)]$$

$$- \int_0^t [1 - F(t+y-x)] \, d E[\alpha(x)]$$

We next wish to show that for any $0 \leqslant u < \infty$,

$$\int_0^u [1 - F(u-x)] \, d E[\alpha(x)] = F(u).$$

To this end we note that

$$E[\alpha(x)] = \sum_{k=1}^{\infty} P[\tau_k \leqslant x] = \sum_{k=1}^{\infty} F_{(k)}(x)$$

where $F_{(1)}(x) = F(x)$ and, for $k \geqslant 1$, $F_{(k+1)}(x)$ is the PDF of τ_{k+1} ($= \tau_k + t_{k+1}$) defined by $F_{(k+1)}(x) = \int_0^x F(x-y) \, d F_{(k)}(y)$. Thus

$$\int_0^u [1 - F(u-x)] \, d E[\alpha(x)] = \int_0^u [1 - F(u-x)] \, d \sum_{k=1}^{\infty} F_{(k)}(x)$$

$$= \sum_{k=1}^{\infty} \int_0^u [1 - F(u-x)] \, d F_{(k)}(x)$$

$$= \sum_{k=1}^{\infty} \left[F_{(k)}(u) - F_{(k+1)}(u) \right]$$

$$= \lim_{k \to \infty} \left[F_{(1)}(u) - F_{(k+1)}(u) \right] = F(u)$$

as desired. Replacing u by $t+y$ in our earlier equation gives

$$F(t+y) = \int_0^{t+y} [1 - F(t+y-x)] \, d E[\alpha(x)]$$

We now apply this result to $\hat{F}_t(y)$ as expressed as the difference of integrals above and obtain

$$\hat{F}_t(y) = F(t+y) - \int_0^t [1 - F(t+y-x)] \, dE[\alpha(x)]$$

and so

$$\hat{F}(y) \triangleq \lim_{t \to \infty} \hat{F}_t(y) = 1 - \lim_{t \to \infty} \int_0^t [1 - F(t+y-x)] \, dE[\alpha(x)]$$

To evaluate this limit we use the key renewal theorem of Smith, which may be found in [FELL 66] pages 347-351 (see also [TAKA 62a] page 227). The key renewal theorem states:
If F is not lattice, and if $h(t)$ is directly Riemann integrable, then

$$\lim_{t \to \infty} \int_0^t h(t-x) \, dE[\alpha(x)] = \frac{1}{m_1} \int_0^\infty h(t) \, dt. \quad \blacksquare$$

We apply this theorem to the function $h(t) = 1 - F(t+y)$ to obtain

$$\lim_{t \to \infty} \int_0^t [1 - F(t+y-x)] \, dE[\alpha(x)] = \frac{1}{m_1} \int_0^\infty [1 - F(t+y)] \, dt.$$

Then

$$\hat{F}(y) = 1 - \frac{1}{m_1} \int_0^\infty [1 - F(t+y)] \, dt$$

$$= \frac{1}{m_1} \left[\int_0^\infty [1 - F(t)] \, dt - \int_y^\infty [1 - F(t)] \, dt \right]$$

$$\hat{F}(y) = \frac{1}{m_1} \int_0^y [1 - F(t)] \, dt$$

and so, finally,

$$\hat{f}(y) = \frac{1 - F(y)}{m_1}.$$

PROBLEM 5.3†

Let us rederive the P−K mean-value formula (5.72).
(a) Recognizing that a new arrival is delayed by one service time for each queued customer plus the residual service time of the customer in service, write an expression for W in terms of $\bar{N}_q$, ρ, $\bar{x}$, and C_b^2.
(b) Use Little's result in (a) to obtain Eq. (5.72).

SOLUTION

(a) Since $r_k = p_k = d_k$ for M/G/1, the average number in queue that an arrival finds is $\bar{N}_q$, the average number in queue (over all time). Thus we may write

$$W = \bar{N}_q \bar{x} + \frac{\overline{x^2}}{2\bar{x}} P[\tilde{w} > 0].$$

But for Poisson arrivals, $P[\tilde{w} > 0] = P[\text{busy system}] = \rho$.

$$\therefore W = \bar{N}_q \bar{x} + \frac{\overline{x^2}}{2\bar{x}} \rho = \bar{N}_q \bar{x} + \frac{\rho \bar{x}}{2(\bar{x})^2}\left[\sigma_b^2 + (\bar{x})^2\right]$$

$$W = \bar{N}_q \bar{x} + \frac{\rho \bar{x}}{2}(1 + C_b^2) \qquad \blacksquare$$

(b) By Little's result, $\bar{N}_q = \lambda W$. So

$$W = \rho W + \frac{\rho \bar{x}}{2}(1 + C_b^2)$$

Thus

$$\frac{W}{\bar{x}} = \frac{\rho(1 + C_b^2)}{2(1-\rho)}$$

which is Eq. (5.72).

PROBLEM 5.4.

Replace $1-\rho$ in Eq. (5.85) by an unknown constant and show that $Q(1) = V(1) = 1$ easily gives us the correct value of $1-\rho$ for this constant.

SOLUTION

We have

$$Q(z) = V(z) \frac{K\left(1 - \frac{1}{z}\right)}{1 - \frac{V(z)}{z}} = K \frac{V(z)(z-1)}{z - V(z)}$$

$$\frac{Q(z)}{V(z)} = K \frac{z-1}{z-V(z)}$$

Since $Q(1) = V(1) = 1$, we may evaluate the constant K by using L'Hospital's rule on the right-hand expression.

$$1 = K \frac{\left.\frac{d}{dz}(z-1)\right|_{z=1}}{\left.\frac{d}{dz}(z-V(z))\right|_{z=1}} = K \frac{1}{1-V^{(1)}(1)}$$

So $K = 1 - V^{(1)}(1) = 1 - \rho$ by Eq. (5.58).

PROBLEM 5.5.†

(a) From Eq. (5.86) form $Q^{(1)}(1)$ and show that it gives the expression for $\bar{q}$ in Eq. (5.63).

(b) From Eq. (5.105), find the first two moments of the waiting time and compare with Eqs. (5.113) and (5.114).

SOLUTION

(a) Eq. (5.86) is $Q(z) = B^*(\lambda - \lambda z) \dfrac{(1-\rho)(1-z)}{B^*(\lambda-\lambda z) - z}$. We first define a function $f(z) = \dfrac{B^*(\lambda - \lambda z) - z}{1 - z}$. This will permit us to factor $(1-z)$ from the denominator of $Q(z)$, thereby eliminating the indeterminacies at $z=1$ and simplifying our calculations. Using the expansion for $B^*(\lambda - \lambda z)$ derived in Exercise (5.1) we see that

$$B^*(\lambda - \lambda z) = 1 - \bar{x}(\lambda - \lambda z) + \frac{\overline{x^2}}{2!}(\lambda - \lambda z)^2 - \frac{\overline{x^3}}{3!}(\lambda - \lambda z)^3 + \cdots$$

and thus

$$f(z) = 1 - \lambda \bar{x} + \frac{\lambda^2 \overline{x^2}}{2!}(1-z) - \frac{\lambda^3 \overline{x^3}}{3!}(1-z)^2 + \cdots$$

So we have $f(1) = 1 - \rho$, $f^{(1)}(1) = -\dfrac{\lambda^2 \overline{x^2}}{2}$. Noting that

$$Q(z) = (1-\rho) \frac{B^*(\lambda - \lambda z)}{f(z)}$$

and hence
$$Q^{(1)}(z) = (1-\rho)\frac{f(z)B^{*(1)}(\lambda - \lambda z)(-\lambda) - B^{*}(\lambda - \lambda z)f^{(1)}(z)}{[f(z)]^2}$$

we obtain
$$\bar{q} = Q^{(1)}(1) = (1-\rho)\left[\frac{(1-\rho)\lambda\bar{x} + \frac{\lambda^2\overline{x^2}}{2}}{(1-\rho)^2}\right] = \rho + \frac{\lambda^2\overline{x^2}}{2(1-\rho)}$$

(which is Eq. (5.62)). Thus
$$\bar{q} = \rho + \rho^2\frac{(1+C_b^2)}{2(1-\rho)}$$

which is Eq. (5.63) as desired.

(b) Eq. (5.105) is $W^{*}(s) = \frac{s(1-\rho)}{s - \lambda + \lambda B^{*}(s)}$. In a fashion similar to part (a), define a function $f(s) = \frac{s - \lambda + \lambda B^{*}(s)}{s} = 1 - \frac{\lambda}{s} + \frac{\lambda}{s}B^{*}(s)$. Again, this permits us to remove the indeterminacy at $s=0$. Using the expansion for $B^{*}(s)$ from Exercise (5.1) we find

$$f(s) = 1 - \frac{\lambda}{s} + \frac{\lambda}{s}\left[1 - s\bar{x} + \frac{s^2\overline{x^2}}{2!} - \frac{s^3\overline{x^3}}{3!} + \frac{s^4\overline{x^4}}{4!} - \cdots\right]$$

$$= 1 - \lambda\bar{x} + \frac{\lambda\overline{x^2}}{2!}s - \frac{\lambda\overline{x^3}}{3!}s^2 + \frac{\lambda\overline{x^4}}{4!}s^3 - \cdots$$

Therefore
$$f^{(1)}(s) = \frac{\lambda\overline{x^2}}{2!} - \frac{2\lambda\overline{x^3}}{3!}s + \frac{3\lambda\overline{x^4}}{4!}s^2 - \cdots$$

and
$$f^{(2)}(s) = -\frac{2\lambda\overline{x^3}}{3!} + \frac{(3\cdot 2)\lambda\overline{x^4}}{4!}s - \cdots$$

Thus we have
$$f(0) = 1 - \rho, \quad f^{(1)}(0) = \frac{\lambda\overline{x^2}}{2}, \quad f^{(2)}(0) = -\frac{\lambda\overline{x^3}}{3}.$$

Recall that $W^{*}(s) = \frac{1-\rho}{f(s)}$. Differentiating,
$$W^{*(1)}(s) = -(1-\rho)\frac{f^{(1)}(s)}{[f(s)]^2}$$

5.5.

and

$$W^{*(2)}(s) = -(1-\rho)\frac{[f(s)]^2 f^{(2)}(s) - 2f(s)[f^{(1)}(s)]^2}{[f(s)]^4}$$

Hence

$$\bar{w} = -W^{*(1)}(0) = \frac{\lambda \overline{x^2}}{2(1-\rho)} = \frac{\lambda b_2}{2(1-\rho)}$$

which is Eq. (5.113) and

$$\overline{w^2} = W^{*(2)}(0) = -(1-\rho)\frac{(1-\rho)^2\left[-\frac{\lambda \overline{x^3}}{3}\right] - 2(1-\rho)\left[\frac{\lambda \overline{x^2}}{2}\right]^2}{(1-\rho)^4}$$

$$\overline{w^2} = 2(\bar{w})^2 + \frac{\lambda \overline{x^3}}{3(1-\rho)} = 2(\bar{w})^2 + \frac{\lambda b_3}{3(1-\rho)}$$

which is Eq. (5.114).

PROBLEM 5.6.†

We wish to prove that the limiting probability r_k for the number of customers found by an arrival is equal to the limiting probability d_k for the number of customers left behind by a departure, in any queueing system in which the state changes by unit step values only (positive or negative). Beginning at $t = 0$, let x_n be those instants when $N(t)$ (the number in system) increases by one and y_n be those instants when $N(t)$ decreases by unity, $n = 1, 2, \ldots$. Let $N(x_n^-)$ be denoted by α_n and $N(y_n^+)$ by β_n. Let $N(0) = i$.
(a) Show that if $\beta_{n+i} \leq k$, then $\alpha_{n+k+1} \leq k$.
(b) Show that if $\alpha_{n+k+1} \leq k$, then $\beta_{n+i} \leq k$.
(c) Show that (a) and (b) must therefore give, for any k,

$$\lim_{n \to \infty} P[\beta_n \leq k] = \lim_{n \to \infty} P[\alpha_n \leq k]$$

which establishes that $r_k = d_k$.

SOLUTION

(a) Assume $\beta_{n+i} \leq k$ (for some $k \geq 0$, $n > 0$); that is, the $(n+i)$th departure leaves behind at most k customers. Let

$a \triangleq$ number of arrivals by time y_{n+i}^+

= number of arrivals prior to the $(n+i)$th departure.

Then $a + i - (n+i) = \beta_{n+i} \leq k$. Hence $a \leq n+k$, and thus the $(n+k+1)$th arrival occurs after the $(n+i)$th departure,

i.e. $y_{n+i}^+ \leq x_{n+k+1}^-$

Since there were at least $n+i$ departures before the $(n+k+1)$th arrival we have

$$\alpha_{n+k+1} \leq n+k+i-(n+i) = k.\qquad\blacksquare$$

(b) Assume $\alpha_{n+k+1} \leq k$ (for some $k \geq 0$, $n > 0$); that is, the $(n+k+1)$th arrival finds at most k customers. Let

$d \triangleq$ number of departures by time x_{n+k+1}^-

= number of departures prior to the $(n+k+1)$th arrival.

Then $n+k+i-d = \alpha_{n+k+1} \leq k$. Hence $n+i \leq d$ and the $(n+i)$th departure must have occurred before the $(n+k+1)$th arrival,

i.e. $y_{n+i}^+ \leq x_{n+k+1}^-$

Since there were at most $n+k$ arrivals by the $(n+i)$th departure we see that

$$\beta_{n+i} \leq n+k+i-(n+i) = k\qquad\blacksquare$$

(c) From parts (a) and (b) (for $k \geq 0$)

$$\beta_{n+i} \leq k \text{ iff } \alpha_{n+k+1} \leq k \quad (n > 0)$$

So

$$P[\beta_{n+i} \leq k] = P[\alpha_{n+k+1} \leq k]$$

Letting $n \to \infty$ (assuming the equilibrium probabilities exist) we have, for fixed $k \geq 0$

$$\lim_{n \to \infty} P[\beta_n \leq k] = \lim_{n \to \infty} P[\beta_{n+i} \leq k]$$

$$= \lim_{n \to \infty} P[\alpha_{n+k+1} \leq k] = \lim_{n \to \infty} P[\alpha_n \leq k]$$

or

$$d_k = r_k \text{ for all } k \geq 0.$$

PROBLEM 5.7.†

In this exercise, we explore the method of supplementary variables as applied to the M/G/1 queue. As usual, let $P_k(t) = P[N(t) = k]$. Moreover, let $P_k(t, x_0)\Delta x_0 = P[N(t) = k, \ x_0 < X_0(t) \leqslant x_0 + \Delta x_0]$ where $X_0(t)$ is the service already received by the customer in service at time t.

(a) Show that
$$\frac{\partial P_0(t)}{\partial t} = -\lambda P_0(t) + \int_0^\infty P_1(t, x_0) r(x_0) \, dx_0$$

where
$$r(x_0) = \frac{b(x_0)}{1 - B(x_0)}$$

(b) Let $p_k = \lim P_k(t)$ as $t \to \infty$ and $p_k(x_0) = \lim P_k(t, x_0)$ as $t \to \infty$. From (a) we have the equilibrium result
$$\lambda p_0 = \int_0^\infty p_1(x_0) r(x_0) \, dx_0$$

Show the following equilibrium results [where $p_0(x_0) \triangleq 0$]:

(i) $\dfrac{\partial p_k(x_0)}{\partial x_0} = -[\lambda + r(x_0)] p_k(x_0) + \lambda p_{k-1}(x_0) \quad k \geqslant 1$

(ii) $p_k(0) = \int_0^\infty p_{k+1}(x_0) r(x_0) \, dx_0 \quad k > 1$

(iii) $p_1(0) = \int_0^\infty p_2(x_0) r(x_0) \, dx_0 + \lambda p_0$

(c) The four equations in (b) determine the equilibrium probabilities when combined with an appropriate normalization equation. In terms of p_0 and $p_k(x_0)$ ($k = 1, 2, \ldots$) give this normalization equation.

(d) Let $R(z, x_0) = \sum_{k=1}^\infty p_k(x_0) z^k$. Show that
$$\frac{\partial R(z, x_0)}{\partial x_0} = [\lambda z - \lambda - r(x_0)] R(z, x_0)$$

and
$$zR(z, 0) = \int_0^\infty r(x_0) R(z, x_0) \, dx_0 + \lambda z(z - 1) p_0$$

(e) Show that the solution for $R(z, x_0)$ from (d) must be
$$R(z, x_0) = R(z, 0) e^{-\lambda x_0 (1 - z) - \int_0^{x_0} r(y) \, dy}$$

5.7.

(f) Defining $R(z) \triangleq \int_0^\infty R(z, x_0) \, dx_0$, show that

$$R(z) = R(z, 0) \frac{1 - B^*(\lambda - \lambda z)}{\lambda(1-z)}$$

$$R(z, 0) = \frac{\lambda z(z-1) p_0}{z - B^*(\lambda - \lambda z)}$$

(g) From the normalization equation of (c), now show that

$$p_0 = 1 - \rho \quad (\rho = \lambda \bar{x})$$

(h) Consistent with Eq. (5.78) we now define

$$Q(z) = p_0 + R(z)$$

Show that $Q(z)$ expressed this way is identical to the P–K transform equation (5.86). (See [COX 55] for additional details of this method.)

SOLUTION

(a) If we are to find zero customers at time $t + \Delta t$, then to within $o(\Delta t)$, it must be that at time t either (i) we had zero customers and there were no arrivals in $(t, t + \Delta t)$ or (ii) we had one customer at time t and this customer completed service in $(t, t + \Delta t)$. We calculate this latter case by conditioning on the service received, $X_0(t)$, and noting that the departure rate when $X_0(t) = x_0$ is simply $r(x_0) = \frac{b(x_0)}{1 - B(x_0)}$. Thus

$$P_0(t + \Delta t) = (1 - \lambda \Delta t) P_0(t) + \int_0^\infty P_1(t, x_0) \, dx_0 \, r(x_0) \, \Delta t + o(\Delta t)$$

Subtracting $P_0(t)$ from both sides, dividing by Δt, and letting $\Delta t \to 0$ gives

$$\frac{\partial P_0(t)}{\partial t} = -\lambda P_0(t) + \int_0^\infty P_1(t, x_0) \, r(x_0) \, dx_0$$

(b) We first note that as $t \to \infty$ then $\frac{\partial P_0(t)}{\partial t} \to 0$, and thus the expression in part (a) becomes

$$\lambda p_0 = \int_0^\infty p_1(x_0) \, r(x_0) \, dx_0.$$

To show equation (i), note that we have a continuous state process; we proceed in a manner analogous to that on page 227. By definition (for $k \geq 1$) $P_k(t + \Delta t, x_0 + \Delta t) \Delta x_0$ is the probability that, at time $t + \Delta t$, there are k in system and the customer being served has already received $x_0 + \Delta t$ seconds of service. The only two ways to enter this state are:

5.7.

(i) there were k in system at time t, the customer in service had received x_0 seconds of service at time t, and there were no arrivals and no departures in $(t, t+\Delta t)$ or

(ii) there were $k-1$ in system at time t, the customer in service had received x_0 seconds of service at time t, and there was one arrival in $(t, t+\Delta t)$ but no departures in $(t, t+\Delta t)$.

Thus

$$P_k(t+\Delta t, x_0+\Delta t)\,\Delta x_0 =$$

$$[1-\lambda\Delta t+o(\Delta t)][1-r(x_0)\Delta t+o(\Delta t)]P_k(t,x_0)\,\Delta x_0$$

$$+[\lambda\Delta t+o(\Delta t)][1-r(x_0)\Delta t+o(\Delta t)]P_{k-1}(t,x_0)\,\Delta x_0$$

Expanding $P_k(t+\Delta t, x_0+\Delta t)$ as

$$P_k(t+\Delta t, x_0+\Delta t) = P_k(t, x_0+\Delta t) + \frac{\partial P_k(t, x_0+\Delta t)}{\partial t}\Delta t + o(\Delta t)$$

we have the equation

$$P_k(t, x_0+\Delta t) - P_k(t, x_0) + \frac{\partial P_k(t, x_0+\Delta t)}{\partial t}\Delta t =$$

$$-[\lambda + r(x_0)]P_k(t, x_0)\Delta t + \lambda P_{k-1}(t, x_0)\Delta t + o(\Delta t)$$

Dividing by Δt and letting $\Delta t \to 0$ gives

$$\frac{\partial P_k(t, x_0)}{\partial x_0} + \frac{\partial P_k(t, x_0)}{\partial t} = -[\lambda + r(x_0)]P_k(t, x_0) + \lambda P_{k-1}(t, x_0)$$

Letting $t \to \infty$ and using the equilibrium result $\lim_{t \to \infty} \frac{\partial P_k(t, x_0)}{\partial t} = 0$ we finally obtain

$$\frac{\partial p_k(x_0)}{\partial x_0} = -[\lambda + r(x_0)]p_k(x_0) + \lambda p_{k-1}(x_0) \quad (k \geq 1)$$

which is equation (i).

We obtain equations (ii) and (iii) as follows. Consider the state in which there are $k \geq 1$ in the system at time $t+\Delta t$ and the customer in service has received Δt seconds of service. For $k > 1$, the only way to enter this state is to have $k+1$ at time t, no arrivals in $(t, t+\Delta t)$, and one departure in $(t, t+\Delta t)$. For $k=1$ there is the added possibility of having an empty system at time t, but one arrival in $(t, t+\Delta t)$. For $k > 1$, we have (conditioning on $X_0(t) = x_0$)

5.7.

$$P_k(t+\Delta t,0)\,\Delta t = \int_0^\infty [1-\lambda\,\Delta t + o(\Delta t)][r(x_0)\,\Delta t + o(\Delta t)]$$
$$\cdot P_{k+1}(t,x_0)\,dx_0$$
$$= \int_0^\infty r(x_0)\,P_{k+1}(t,x_0)\,dx_0\,\Delta t + o(\Delta t)$$

Dividing by Δt and letting $\Delta t \to 0$ gives
$$P_k(t,0) = \int_0^\infty P_{k+1}(t,x_0)\,r(x_0)\,dx_0.$$

Letting $t \to \infty$ we finally have
$$p_k(0) = \int_0^\infty p_{k+1}(x_0)\,r(x_0)\,dx_0 \quad (k>1)$$

For $k=1$ a similar argument (we may now have an arrival to an empty system) gives
$$p_1(0) = \int_0^\infty p_1(x_0)\,r(x_0)\,dx_0 + \lambda\,p_0.$$

(c) For $k \geq 1$ we clearly have $p_k = \int_0^\infty p_k(x_0)\,dx_0$. Thus the normalization $\sum_{k=0}^\infty p_k = 1$ becomes
$$p_0 + \sum_{k=1}^\infty \int_0^\infty p_k(x_0)\,dx_0 = 1. \quad\blacksquare$$

(d) Multiply each equation in (i) of part (b) by z^k and sum for $k \geq 1$. Then
$$\sum_{k=1}^\infty \frac{\partial p_k(x_0)}{\partial x_0} z^k = -[\lambda + r(x_0)]\sum_{k=1}^\infty p_k(x_0)\,z^k + \lambda\sum_{k=1}^\infty p_{k-1}(x_0)\,z^k.$$

Recalling that $p_0(x_0) = 0$ and $R(z,x_0) = \sum_{k=1}^\infty p_k(x_0)\,z^k$ we find
$$\frac{\partial R(z,x_0)}{\partial x_0} = [\lambda z - \lambda - r(x_0)]\,R(z,x_0).$$

Next, using (ii) and (iii) of part (b),
$$R(z,0) = \sum_{k=1}^\infty p_k(0)\,z^k$$
$$= \lambda p_0 z + \sum_{k=1}^\infty \int_0^\infty p_{k+1}(x_0)\,r(x_0)\,dx_0\,z^k$$
$$= \lambda p_0 z + \int_0^\infty \left[\sum_{k=1}^\infty p_{k+1}(x_0)\,z^k\right] r(x_0)\,dx_0$$

5.7.

Using the definition of $R(z, x_0)$ we finally have
$$R(z,0) = \lambda p_0 z + \int_0^\infty \frac{1}{z} [R(z, x_0) - p_1(x_0) z] \, r(x_0) \, dx_0$$
$$= \lambda p_0 z + \frac{1}{z} \int_0^\infty R(z, x_0) \, r(x_0) \, dx_0 - \lambda p_0$$

or
$$z R(z,0) = \lambda p_0 z(z-1) + \int_0^\infty R(z, x_0) \, r(x_0) \, dx_0.$$

(e) From part (d) we have
$$\frac{\frac{\partial R(z, y)}{\partial y}}{R(z, y)} = \lambda z - \lambda - r(y)$$

Integrate from 0 to x_0 to obtain
$$\log_e[R(z, x_0)] - \log_e[R(z, 0)] = [\lambda z - \lambda] x_0 - \int_0^{x_0} r(y) \, dy$$

or
$$R(z, x_0) = R(z, 0) \, e^{\lambda(z-1) x_0 - \int_0^{x_0} r(y) \, dy}$$

as desired. We next note that
$$e^{-\int_0^{x_0} r(y) \, dy} = e^{-\int_0^{x_0} \frac{b(y)}{1 - B(y)} \, dy} = e^{\log_e[1 - B(x_0)]} = 1 - B(x_0)$$

So we can rewrite our previous result as
$$R(z, x_0) = R(z, 0) \, e^{\lambda(z-1) x_0}[1 - B(x_0)].$$

Using this expression in the equation for $z R(z, 0)$ obtained in part (d) yields
$$z R(z,0) = \lambda p_0 z(z-1) + \int_0^\infty R(z, 0) \, e^{\lambda(z-1) x_0}[1 - B(x_0)] \cdot r(x_0) \, dx_0$$

or
$$z R(z,0) = \lambda p_0 z(z-1) + R(z, 0) \int_0^\infty e^{\lambda(z-1) x_0} b(x_0) \, dx_0$$
$$= \lambda p_0 z(z-1) + R(z, 0) \, B^*(\lambda - \lambda z)$$

$$\therefore R(z,0) = \frac{\lambda z(z-1) p_0}{z - B^*(\lambda - \lambda z)}$$

5.7.

(f)
$$R(z) \triangleq \int_0^\infty R(z, x_0) \, dx_0$$
$$= \int_0^\infty R(z,0) \, e^{\lambda(z-1)x_0}[1 - B(x_0)] \, dx_0$$
$$= R(z,0) \int_0^\infty e^{-\lambda(1-z)x_0}[1 - B(x_0)] \, dx_0$$

Recalling the steps from Eq. (5.10) to Eq. (5.11) we have
$$R(z) = R(z,0) \frac{1 - B^*(\lambda - \lambda z)}{\lambda(1 - z)}$$

(g) From the normalization equation $p_0 + \int_0^\infty \sum_{k=1}^\infty p_k(x_0) \, dx_0 = 1$ and the definition of $R(1, x_0)$ we have $p_0 + \int_0^\infty R(1, x_0) \, dx_0 = 1$. Thus
$$p_0 + R(1) = 1.$$

From (f) and (e) we also have
$$R(z) = R(z,0) \frac{1 - B^*(\lambda - \lambda z)}{\lambda(1 - z)}$$
$$= \frac{\lambda z(z-1)p_0}{z - B^*(\lambda - \lambda z)} \cdot \frac{1 - B^*(\lambda - \lambda z)}{\lambda(1 - z)}$$

or
$$\frac{R(z)}{z} = p_0 \frac{1 - B^*(\lambda - \lambda z)}{B^*(\lambda - \lambda z) - z}$$

Letting $z \to 1$ (and using L'Hospital's rule) we find
$$R(1) = p_0 \lim_{z \to 1} \frac{1 - B^*(\lambda - \lambda z)}{B^*(\lambda - \lambda z) - z}$$
$$= p_0 \left. \frac{-B^{*(1)}(\lambda - \lambda z)(-\lambda)}{B^{*(1)}(\lambda - \lambda z)(-\lambda) - 1} \right|_{z=1}$$
$$= p_0 \frac{-\lambda \bar{x}}{\lambda \bar{x} - 1} = p_0 \frac{\rho}{1 - \rho}$$

But then $1 = p_0 + R(1) = p_0 + p_0 \frac{\rho}{1-\rho}$ and so
$$p_0 = 1 - \rho.$$

5.7.

(h)

$$Q(z) = p_0 + R(z)$$

$$= p_0 + \frac{p_0 z[1 - B^*(\lambda - \lambda z)]}{B^*(\lambda - \lambda z) - z}$$

$$= p_0 \frac{B^*(\lambda - \lambda z)(1 - z)}{B^*(\lambda - \lambda z) - z}$$

So

$$Q(z) = B^*(\lambda - \lambda z) \frac{(1 - \rho)(1 - z)}{B^*(\lambda - \lambda z) - z}$$

which is Eq. (5.86).

PROBLEM 5.8.

Consider the M/G/∞ queue in which each customer always finds a free server; thus

$$s(y) = b(y) \text{ and } T = \bar{x}. \text{ Let } P_k(t) = P[N(t) = k] \qquad \blacksquare$$

and assume $P_0(0) = 1$.

(a) Show that

$$P_k(t) = \sum_{n=k}^{\infty} e^{-\lambda t} \frac{(\lambda t)^n}{n!} \binom{n}{k} \left[\frac{1}{t} \int_0^t [1 - B(x)] \, dx \right]^k \left[\frac{1}{t} \int_0^t B(x) \, dx \right]^{n-k}$$

[HINT: $(1/t) \int_0^t B(x) \, dx$ is the probability that a customer's service terminates by time t, given that his arrival time was uniformly distributed over the interval $(0, t)$. See Eq. (2.137) also.]

(b) Show that $p_k \triangleq \lim P_k(t)$ as $t \to \infty$ is

$$p_k = \frac{(\lambda \bar{x})^k}{k!} e^{-\lambda \bar{x}} \qquad \blacksquare$$

regardless of the form of $B(x)$!

SOLUTION

(a) We determine the distribution of $N(t)$, the number of customers in system at time t, by conditioning on the number of arrivals in $(0, t)$. Thus

$$P_k(t) = P[N(t) = k]$$

$$= \sum_{n=0}^{\infty} P[N(t) = k \mid n \text{ arrivals in } (0, t)] \cdot P[n \text{ arrivals in } (0, t)]$$

$$= \sum_{n=0}^{\infty} P[N(t) = k \mid n \text{ arrivals in } (0, t)] \cdot \frac{e^{-\lambda t}(\lambda t)^n}{n!}$$

The probability that a customer who arrived at time x ($0 < x < t$) is still in the system at time t is simply $1 - B(t - x)$. By Eq. (2.137), since arrivals form a Poisson process, the joint distribution of the arrival times given that n arrivals occurred in $(0, t)$ is the same as the distribution of n points uniformly distributed over $(0, t)$. Thus for any number n of arrivals in $(0, t)$, the probability of any customer still being present at time t is $\int_0^t [1 - B(t - x)] \frac{dx}{t} = \frac{1}{t} \int_0^t [1 - B(x)] \, dx$. Hence the probability that a customer who arrived in $(0, t)$ will have completed service by time t is $1 - \frac{1}{t} \int_0^t [1 - B(x)] \, dx = \frac{1}{t} \int_0^t B(x) \, dx$. Thus

$$P[N(t) = k \mid n \text{ arrivals in } (0, t)] =$$

$$\binom{n}{k} \left[\frac{1}{t} \int_0^t [1 - B(x)] \, dx \right]^k \left[\frac{1}{t} \int_0^t B(x) \, dx \right]^{n-k}$$

for $k \leq n$ (and $= 0$ for $k > n$). So

$$P_k(t) = \sum_{n=k}^{\infty} \binom{n}{k} \left[\frac{1}{t} \int_0^t [1 - B(x)] \, dx \right]^k \left[\frac{1}{t} \int_0^t B(x) \, dx \right]^{n-k} \cdot \frac{e^{-\lambda t}(\lambda t)^n}{n!}$$

which is the desired result.

(b) We first simplify the expression obtained in part (a), and then find the limit as $t \to \infty$.

$$P[N(t) = k] = \sum_{n=k}^{\infty} e^{-\lambda t} \frac{(\lambda t)^n}{n!} \binom{n}{k} \left[\frac{1}{t} \int_0^t [1 - B(x)] \, dx \right]^k \left[\frac{1}{t} \int_0^t B(x) \, dx \right]^{n-k}$$

$$= e^{-\lambda t} \sum_{n=k}^{\infty} \frac{\left[\lambda \int_0^t [1 - B(x)] \, dx \right]^k}{k!} \cdot \frac{\left[\lambda \int_0^t B(x) \, dx \right]^{n-k}}{(n-k)!}$$

$$= e^{-\lambda t} \frac{\left[\lambda \int_0^t [1 - B(x)] \, dx \right]^k}{k!} \sum_{n=0}^{\infty} \frac{\left[\lambda \int_0^t B(x) \, dx \right]^n}{n!}$$

5.8.

$$P[N(t) = k] = e^{-\lambda t}\frac{\left[\lambda\int_0^t [1-B(x)]\,dx\right]^k}{k!} e^{\lambda\int_0^t B(x)\,dx}$$

$$= e^{-\lambda\int_0^t [1-B(x)]\,dx}\frac{\left[\lambda\int_0^t [1-B(x)]\,dx\right]^k}{k!} \quad\blacksquare$$

(Thus we may note that, *for every* t, $N(t)$ is Poisson with parameter $\lambda\int_0^t [1-B(x)]\,dx$.) Letting $t\to\infty$ and noting that

$$\lim_{t\to\infty}\int_0^t [1-B(x)]\,dx = \int_0^\infty [1-B(x)]\,dx = \bar{x}$$

we see immediately that

$$p_k \triangleq \lim_{t\to\infty} P_k(t) = e^{-\lambda\bar{x}}\frac{(\lambda\bar{x})^k}{k!}.$$

Thus as $t\to\infty$, the limiting distribution of number in system is Poisson with parameter $\lambda\bar{x}$ which is independent (except for the mean) of $B(x)$.

PROBLEM 5.9.

Consider M/E$_2$/1.
(a) Find the polynomial for $G^*(s)$.
(b) Solve for $S(y) = P[\text{time in system} \leq y]$.

SOLUTION

(a) For the M/E$_2$/1 system, the Laplace transform of the service time density is $B^*(s) = \left(\dfrac{2\mu}{s+2\mu}\right)^2$. Thus Eq. (5.137) gives

$$G^*(s) = \left[\frac{2\mu}{s+\lambda-\lambda G^*(s)+2\mu}\right]^2.$$

Expanding, we get

$$\lambda^2[G^*(s)]^3 - 2\lambda(s+\lambda+2\mu)[G^*(s)]^2 + (s+\lambda+2\mu)^2 G^*(s) - 4\mu^2 = 0 \quad\blacksquare$$

(b) Eq. (5.100) gives

$$S^*(s) = B^*(s)\frac{s(1-\rho)}{s-\lambda+\lambda B^*(s)}$$

Thus

$$S^*(s) = \left(\frac{2\mu}{s+2\mu}\right)^2 \frac{s(1-\rho)}{s-\lambda+\lambda\left(\frac{2\mu}{s+2\mu}\right)^2}$$

$$= \frac{4\mu^2(1-\rho)}{s^2+(4\mu-\lambda)s+4\mu(\mu-\lambda)}$$

The denominator $s^2+(4\mu-\lambda)s+4\mu(\mu-\lambda)$ has roots s_1, s_2 (where $\rho = \lambda/\mu$):

$$s_1 = \frac{-\mu(4-\rho)+\mu\sqrt{\rho^2+8\rho}}{2}$$

$$s_2 = \frac{-\mu(4-\rho)-\mu\sqrt{\rho^2+8\rho}}{2}$$

We note that, for $\rho < 1$, we have $16\rho < 16$ and thus $(4-\rho)^2 > \rho^2+8\rho$. Hence $s_2 < s_1 < 0$ for $0 < \rho < 1$. Factoring,

$$S^*(s) = \frac{4\mu^2(1-\rho)}{(s-s_1)(s-s_2)}$$

$$= \frac{4\mu^2(1-\rho)}{\mu\sqrt{\rho^2+8\rho}}\left[\frac{1}{s-s_1}-\frac{1}{s-s_2}\right]$$

Invert to find the pdf $s(y)$ as

$$s(y) = \frac{4\mu(1-\rho)}{\sqrt{\rho^2+8\rho}}\left[e^{s_1 y}-e^{s_2 y}\right]$$

Thus the PDF $S(y)$ is

$$S(y) = \frac{4\mu(1-\rho)}{\sqrt{\rho^2+8\rho}}\left[\frac{1}{s_1}\left(e^{s_1 y}-1\right)-\frac{1}{s_2}\left(e^{s_2 y}-1\right)\right]$$

PROBLEM 5.10.

Consider an M/D/1 system for which $\bar{x} = 2$ sec.
(a) Show that the residual service time pdf $\hat{b}(x)$ is a rectangular distribution.
(b) For $\rho = 0.25$, show that the result of Eq. (5.111) with four terms may be used as a good approximation to the distribution of queueing time.

SOLUTION

(a) The service time distribution is given by
$$B(x) = \begin{cases} 0 & x < 2 \\ 1 & x \geq 2 \end{cases}$$

The residual service time pdf is
$$\hat{b}(x) = \frac{1 - B(x)}{\bar{x}} = \begin{cases} \frac{1}{2} & x < 2 \\ 0 & x \geq 2 \end{cases}$$

Thus $\hat{b}(x)$ is rectangular.

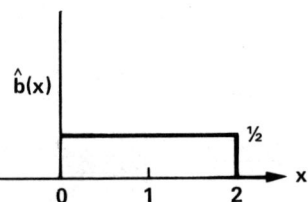

(b) Eq. (5.11) gives $w(y) = \sum_{k=0}^{\infty} (1-\rho)\rho^k \hat{b}_{(k)}(y)$ where $\hat{b}_0(y) = u_0(y)$ and for $k \geq 1$, $\hat{b}_k(y)$ is the k-fold convolution of the pdf $\hat{b}(y)$ with itself. The first four terms of this series give
$$w(y) \cong w_{approx}(y) \triangleq (1-\rho)\left[u_0(y) + \rho\hat{b}(y) + \rho^2\hat{b}_{(2)}(y) + \rho^3\hat{b}_{(3)}(y)\right]$$

As
$$\hat{b}(y) = \begin{cases} \frac{1}{2} & y < 2 \\ 0 & y \geq 2 \end{cases}$$

we see that
$$\hat{b}_{(2)}(y) = \begin{cases} \frac{y}{4} & 0 \leq y \leq 2 \\ 1 - \frac{y}{4} & 2 \leq y \leq 4 \end{cases}$$

5.10.

and

$$\hat{b}_{(3)}(y) = \begin{cases} \dfrac{y^2}{16} & 0 \leq y \leq 2 \\ \dfrac{-y^2}{8} + \dfrac{3}{4}y - \dfrac{3}{4} & 2 \leq y \leq 4 \\ \dfrac{y^2}{16} - \dfrac{3}{4}y + \dfrac{9}{4} & 4 \leq y \leq 6 \end{cases}$$

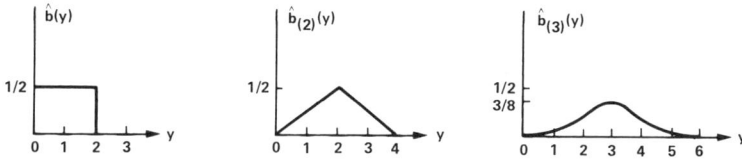

We compare $w(y)$ and $w_{approx}(y)$ in three different ways.

First, the area A under the curve $w(y)$ minus the area A_{approx} under the curve $w_{approx}(y)$ is

$$A - A_{approx} = (1-\rho)\sum_{k=4}^{\infty}\rho^k \int_0^{\infty} \hat{b}_{(k)}(y)\,dy = (1-\rho)\sum_{k=4}^{\infty}\rho^k$$

$$= (1-\rho)\rho^4\left[\dfrac{1}{1-\rho}\right] = \rho^4$$

As $\rho = \dfrac{1}{4}$,

$$A - A_{approx} = \dfrac{1}{256}.$$

Thus, in terms of area, we have a "good" approximation.

Second, we note that $w_{approx}(y) = 0$ for $y \geq 6$. Thus the tail of the density $w(y)$ is *not* approximated very well.

Third, we compare the mean wait W with an approximation W_{approx} calculated from $w_{approx}(y)$. (Note that $w_{approx}(y)$ is not a pdf.)

$$W = \int_0^{\infty} y\,w(y)\,dy = (1-\rho)\sum_{k=1}^{\infty}\rho^k \int_0^{\infty} y\,\hat{b}_{(k)}(y)\,dy$$

We now observe that $\int_0^{\infty} y\,\hat{b}_{(k)}(y)\,dy$ has value k, since it represents the mean of a sum of k random variables each having mean 1. Thus

5.10.

$$W = (1-\rho)\sum_{k=1}^{\infty} k\rho^k = (1-\rho)\rho\frac{\partial}{\partial\rho}\left(\frac{1}{1-\rho}\right)$$

or

$$W = \frac{\rho}{1-\rho}$$

For $\rho = \frac{1}{4}$,

$$W = \frac{1}{3}$$

Now

$$W_{approx} = \int_0^\infty y\, w_{approx}(y)\, dy = (1-\rho)\sum_{k=1}^{3}\rho^k \int_0^\infty y\, \hat{b}_{(k)}(y)\, dy$$

$$= (1-\rho)\sum_{k=1}^{3} k\rho^k = (1-\rho)(\rho + 2\rho^2 + 3\rho^3)$$

For $\rho = \frac{1}{4}$,

$$W_{approx} = \frac{3}{4}\left(\frac{1}{4} + \frac{1}{8} + \frac{3}{64}\right) = \frac{3}{4}\cdot\frac{27}{64} = \frac{81}{256}$$

or

$$W_{approx} = .31640625$$

Thus

$$\frac{W - W_{approx}}{W} = \frac{\frac{1}{3} - \left(\frac{3}{4}\right)^4}{\frac{1}{3}} \cong .0508$$

and so W_{approx} is within 5% of the mean W.

PROBLEM 5.11.

Consider an M/G/1 queue in which bulk arrivals occur at rate λ and with a probability g_r that r customers arrive together at an arrival instant.
(a) Show that the z-transform of the number of customers arriving in an interval of length t is $e^{-\lambda t[1-G(z)]}$ where $G(z) = \sum g_r z^r$.
(b) Show that the z-transform of the random variables v_n, the number of arrivals during the service of a customer, is $B^*[\lambda - \lambda G(z)]$.

5.10. – 5.11.

SOLUTION

(a) Let the random variable $N = N(t)$ represent the number of bulks that arrive in an interval of length t, and let $N_1(z)$ be its z-transform. The total number of customers to arrive is $Y = X_1 + \cdots + X_N$ where X_i is the size of the ith bulk. Then, as in Eq. (II.34), we may write down the z-transform $Y(z)$ for number of customers as $Y(z) = N_1(G(z))$. This result may be derived directly by conditioning on N. Thus

$$Y(z \mid N = n) \triangleq E[z^Y \mid N = n] = E[z^{X_1 + \cdots + X_n}]$$

$$= E[z^{X_1}] \cdots E[z^{X_n}] = [G(z)]^n.$$

Unconditioning gives

$$Y(z) = \sum_{n=0}^{\infty} Y(z \mid N = n) \, P[N = n]$$

$$= \sum_{n=0}^{\infty} [G(z)]^n P[N = n] = N_1(G(z))$$

as above. But $N_1(z) = e^{-\lambda t(1-z)}$ since $N = N(t)$ is Poisson with parameter λt. Thus we have

$$Y(z) = N_1(G(z)) = e^{-\lambda t[1 - G(z)]}.$$

(b)

$$V(z) = E[z^{\tilde{v}}] = \sum_{k=0}^{\infty} P[\tilde{v} = k] z^k$$

$$= \sum_{k=0}^{\infty} \int_0^{\infty} P[\tilde{v} = k \mid \tilde{x} = x] \, b(x) \, dx \, z^k$$

$$= \int_0^{\infty} \sum_{k=0}^{\infty} P[\tilde{v} = k \mid \tilde{x} = x] \, z^k \, b(x) \, dx$$

By part (a) above, $\sum_{k=0}^{\infty} P[\tilde{v} = k \mid \tilde{x} = x] \, z^k = e^{-\lambda x[1 - G(z)]}$. (It is the z-transform of the number of customers to arrive during an interval of length x.) Thus

$$V(z) = \int_0^{\infty} e^{-\lambda x[1 - G(z)]} b(x) \, dx$$

$$V(z) = B^*[\lambda - \lambda G(z)].$$

PROBLEM 5.12.†

Consider the M/G/1 bulk arrival system in the previous problem. Using the method of imbedded Markov chains:

(a) Find the expected queue size at departure instants.
  [HINT: show that $\bar{v} = \rho = \lambda \bar{x} \bar{g}$ and

$$\overline{v^2} - \bar{v} = \left. \frac{d^2 V(z)}{dz^2} \right|_{z=1} = \rho^2(1+C_b^2) + \rho \bar{g}(1+C_g^2) - \rho$$

  where C_g is the coefficient of variation of the bulk group size and $\bar{g}$ is the mean group size.]

(b) Show that the generating function for queue size at departure instants is

$$Q(z) = \frac{(1-\rho)[1-G(z)]B^*[\lambda - \lambda G(z)]}{\bar{g}(B^*[\lambda - \lambda G(z)] - z)} \quad \blacksquare$$

(c) Using the same method (imbedded Markov chain) find the expected number of groups in the queue (averaged over departure times). [HINTS: Show that $D(z) = \beta^*(\lambda - \lambda z)$, where $D(z)$ is the generating function for the number of *groups* arriving during the service time for an entire *group* and where $\beta^*(s)$ is the Laplace transform of the service-time density for an entire group. Also note that $\beta^*(s) = G[B^*(s)]$, which allows us to show that $\tau^2 = (\bar{x})^2(\overline{g^2} - \bar{g}) + \overline{x^2}\bar{g}$, where τ^2 is the second moment of the *group* service time.]

(d) Using Little's result, find W_g, the expected wait on queue for a group (measured from the arrival time of the group until the start of service of the *first* member of the group) and show that

$$W_g = \frac{\rho \bar{x} \bar{g}}{2(1-\rho)} \left[1 + \frac{C_b^2}{\bar{g}} + C_g^2\right]$$

(e) If the customers within a group arriving together are served in random order, show that the ratio of the mean waiting time for a single customer to the average service time for a single customer is $W_g/\bar{x}$ from (d) increased by $(1/2)\bar{g}(1+C_g^2) - 1/2$.

SOLUTION

(a) We begin by finding the first two moments of the random variable $\tilde{v}$ (which has the same distribution as v_n) as suggested in the hint. By Exercise 5.11(b) $V(z) = B^*[\lambda - \lambda G(z)]$. Thus

5.12.

$$V^{(1)}(z) = B^{*(1)}[\lambda - \lambda G(z)][-\lambda G^{(1)}(z)]$$

and

$$V^{(2)}(z) = B^{*(2)}[\lambda - \lambda G(z)][\lambda G^{(1)}(z)]^2 + B^{*(1)}[\lambda - \lambda G(z)][-\lambda G^{(2)}(z)].$$

Hence $\bar{v} = V^{(1)}(1) = B^{*(1)}(0)[-\lambda G^{(1)}(1)] = \lambda \bar{g} \bar{x}$. Since the average arrival rate is $\lambda \bar{g}$ we have

$$\bar{v} = \lambda \bar{g} \bar{x} = \rho.$$

Also

$$\overline{v^2} - \bar{v} = V^{(2)}(1) = B^{*(2)}(0)[\lambda G^{(1)}(1)]^2 + B^{*(1)}(0)[-\lambda G^{(2)}(1)]$$

$$= \overline{x^2}(\lambda \bar{g})^2 + \lambda \bar{x}(\overline{g^2} - \bar{g})$$

$$= (\lambda \bar{g} \bar{x})^2 \frac{\overline{x^2}}{(\bar{x})^2} + \lambda \bar{x} \bar{g} \left[\frac{\overline{g^2}}{\bar{g}} - 1\right]$$

$$= \rho^2(1 + C_b^2) + \rho \bar{g}(1 + C_g^2) - \rho$$

Thus

$$\overline{v^2} = \rho^2(1 + C_b^2) + \rho \bar{g}(1 + C_g^2).$$

We next derive an equation for q_{n+1}, the number left behind by a departure. We first note that if $q_n > 0$, then clearly $q_{n+1} = q_n - 1 + v_{n+1}$ where $\tilde{v}$ has the z-transform $V(z) = B^*[\lambda - \lambda G(z)]$. However, if $q_n = 0$ the next customer to arrive to this empty system will be in a bulk whose size is a random variable, say $\tilde{g}$; thus $q_{n+1} = \tilde{g} - 1 + v_{n+1}$. Hence

$$q_{n+1} = \begin{cases} v_{n+1} - 1 + q_n & q_n > 0 \\ v_{n+1} - 1 + \tilde{g} & q_n = 0 \end{cases}$$

where $\tilde{g}$ has z-transform $G(z)$. Define a new random variable

$$\Delta_{\tilde{g}, k} \triangleq \begin{cases} k & k > 0 \\ \tilde{g} & k = 0 \end{cases}$$

Thus $q_{n+1} = v_{n+1} - 1 + \Delta_{\tilde{g}, q_n}$. Take expectations and let $n \to \infty$. Then $\bar{q} = \bar{v} - 1 + E[\Delta_{\tilde{g}, \tilde{q}}]$. But

$$E[\Delta_{\tilde{g}, \tilde{q}}] = E[\tilde{g}] P[\tilde{q} = 0] + \sum_{k=1}^{\infty} k P[\tilde{q} = k]$$

5.12.

and so
$$E\left[\Delta_{\tilde{g},\tilde{q}}\right] = \bar{g}\,d_0 + \bar{q}$$

Therefore
$$\bar{q} = \bar{v} - 1 + \bar{g}\,d_0 + \bar{q}$$

or
$$d_0 = \frac{1-\bar{v}}{\bar{g}} = \frac{1-\rho}{\bar{g}}.$$

To find $\overline{q^2}$, we must square our equation and try again.
$$q_{n+1}^2 = (v_{n+1}-1)^2 + 2(v_{n+1}-1)\Delta_{\tilde{g},q_n} + \Delta_{\tilde{g},q_n}^2$$

Taking expectations, using independence, and letting $n \to \infty$, we have
$$\overline{q^2} = \overline{v^2} - 2\bar{v} + 1 + 2(\bar{v}-1)E\left[\Delta_{\tilde{g},\tilde{q}}\right] + E\left[\Delta_{\tilde{g},\tilde{q}}^2\right]$$

But
$$E\left[\Delta_{\tilde{g},\tilde{q}}^2\right] = E[\tilde{g}^2]\,P[\tilde{q}=0] + \sum_{k=1}^{\infty} k^2\,P[\tilde{q}=k]$$
$$= \overline{g^2}\,d_0 + \overline{q^2}$$

So
$$\overline{q^2} = \overline{v^2} - 2\bar{v} + 1 + 2(\bar{v}-1)(\bar{g}\,d_0 + \bar{q}) + \overline{g^2}\,d_0 + \overline{q^2}$$

Substituting for $\overline{v^2}$, $\bar{v}$ and d_0 we have
$$0 = \rho^2(1+C_b^2) + \rho\,\bar{g}(1+C_g^2) - 2\rho + 1 + 2(\rho-1)(1-\rho+\bar{q}) + \frac{\overline{g^2}}{\bar{g}}(1-\rho)$$

or
$$2(1-\rho)\bar{q} = \rho^2(1+C_b^2) + \rho\,\bar{g}(1+C_g^2) + 2(1-\rho) - 1 - 2(1-\rho)^2$$
$$+ \bar{g}(1+C_g^2)(1-\rho)$$

and so
$$2(1-\rho)\bar{q} = \rho^2(1+C_b^2) + \bar{g}(1+C_g^2) + 2(1-\rho)\rho - 1$$

$$\therefore \bar{q} = \rho + \frac{\rho^2(1+C_b^2)}{2(1-\rho)} + \frac{\bar{g}(1+C_g^2) - 1}{2(1-\rho)} \qquad \blacksquare$$

5.12.

(b)

$$Q_{n+1}(z) = E\left[z^{q_{n+1}}\right] = E\left[z^{v_{n+1}-1+\Delta_{\tilde{g},q_n}^-}\right]$$

$$= E\left[z^{v_{n+1}-1}\right] \cdot E\left[z^{\Delta_{\tilde{g},q_n}^-}\right]$$

by independence of v_{n+1}, $\tilde{g}$, q_n. Let $n \to \infty$ and obtain

$$Q(z) = \frac{V(z)}{z} E\left[z^{\Delta_{\tilde{g},\tilde{q}}^-}\right]$$

But

$$E\left[z^{\Delta_{\tilde{g},\tilde{q}}^-}\right] = E[z^{\tilde{g}}] \cdot P[\tilde{q}=0] + \sum_{k=1}^{\infty} z^k P[\tilde{q}=k]$$

$$= G(z)\, d_0 + [Q(z) - d_0]$$

Thus

$$Q(z) = \frac{V(z)}{z}[Q(z) + [G(z)-1]d_0]$$

$$Q(z) = V(z)\frac{d_0[1-G(z)]}{V(z)-z}$$

But $V(z) = B^*[\lambda - \lambda G(z)]$ and $d_0 = \dfrac{1-\rho}{\bar{g}}$.

$$\therefore\ Q(z) = B^*[\lambda - \lambda G(z)]\frac{(1-\rho)[1-G(z)]}{\bar{g}\,(B^*[\lambda-\lambda G(z)]-z)} \quad \blacksquare$$

(c) Since we seek the expected number of groups in the queue (averaged over departure instants), we study an M/G/1 queue where the nth group is considered to be the nth "customer" whose service time corresponds to that for the entire nth group. Recognizing this, we may apply the results of Chapter 5 directly — only the arrival and service processes must first be determined. Thus Eq. (5.46) immediately gives $D(z) = \beta^*(\lambda - \lambda z)$. Also, clearly by Eq. (II.34) in Appendix II, we have $\beta^*(s) = G[B^*(s)]$. (The nth group's service time is composed of a sum of a random number — the bulk size — of individual service times.) So by Eq. (5.62) and Eq. (5.65) we have:

$$\overline{D_q} \triangleq \text{expected number of groups in queue}$$

satisfies

5.12.

$$\overline{D_q} = \frac{\lambda^2 \overline{\tau^2}}{2(1-\lambda \overline{\tau})}$$

where $\overline{\tau}$ and $\overline{\tau^2}$ are the first two moments of the group service time. Note that

$$\overline{\tau} = -\beta^{*(1)}(0) = \overline{g}\,\overline{x} \quad \text{(see p 388)}$$

(the perfectly reasonable result that the mean group service time is just $\overline{x}$ times the mean bulk size) and

$$\overline{\tau^2} = \beta^{*(2)}(0) = \overline{g}\,\overline{x^2} + (\overline{g^2} - \overline{g})(\overline{x})^2 \quad \text{(see p 388)}$$

We may use these results and the fact that $\rho = \lambda \overline{\tau}$ to obtain

$$\overline{D_q} = \frac{\lambda^2[\overline{g}\,\overline{x^2} + (\overline{g^2} - \overline{g})(\overline{x})^2]}{2(1-\rho)}$$

$$\overline{D_q} = \frac{(\lambda \overline{g}\,\overline{x})^2}{2(1-\rho)}\left[\frac{\overline{g^2}}{\overline{g}^2} + \left(\frac{\overline{x^2} - (\overline{x})^2}{\overline{g}\,(\overline{x})^2}\right)\right]$$

$$\overline{D_q} = \frac{\rho^2}{2(1-\rho)}\left[1 + C_g^2 + \frac{C_b^2}{\overline{g}}\right] \quad \blacksquare$$

(d) By Little's result (applied on the "group" queue)

$$\overline{D_q} = \lambda W_g$$

Hence

$$W_g = \frac{\rho \overline{g}\,\overline{x}}{2(1-\rho)}\left[1 + C_g^2 + \frac{C_b^2}{\overline{g}}\right].$$

(e) The waiting time for a customer consists of the waiting time until someone in his group gets served (expected value W_g was found in part (d)) plus the waiting time due to service of members within his group (call the expected value of this waiting time W_s). Pick a random customer. Given that this customer came from a bulk of size r, his expected wait due to people in this bulk is

$$\frac{1}{r}\left[0 + \overline{x} + 2\overline{x} + \cdots + (r-1)\overline{x}\right] = \frac{1}{r} \cdot \frac{r(r-1)}{2}\overline{x} = \frac{r-1}{2}\overline{x}.$$

We now must find the probability, say $\hat{p}_r$, that this random customer came from a bulk of size r. To this end, consider an arbitrarily long time interval τ. Bulks of size r arrive at a rate λg_r, each bringing r customers. Thus an average of $\lambda g_r r \tau$ customers from bulks of size r will arrive during τ. Similarly the total number of customers to arrive during τ will be, on the average, simply $\sum_{k=1}^{\infty} \lambda g_k k \tau$. Thus the probability $\hat{p}_r$ that a random

5.12.

customer will be in a bulk of size r is approximated by

$$\hat{p}_r \cong \frac{\lambda g_r r \tau}{\sum_{k=1}^{\infty} \lambda g_k k \tau}.$$

Letting $\tau \to \infty$ we obtain

$$\hat{p}_r = P[\text{random customer was in bulk of size } r]$$

$$= \frac{r g_r}{\sum_{k=1}^{\infty} k g_k} = \frac{r g_r}{\bar{g}}.$$

[Note: this type of argument extends to the continuous variable case.] So

$$W_s = \sum_{r=1}^{\infty} (W_s | \text{customer from bulk of size } r) \cdot \hat{p}_r$$

$$= \sum_{r=1}^{\infty} \left(\frac{r-1}{2} \bar{x}\right) \frac{r g_r}{\bar{g}}$$

$$W_s = \frac{\bar{x}}{2\bar{g}} \left[\sum_{r=1}^{\infty} (r^2 - r) g_r\right] = \frac{\bar{x}}{2\bar{g}} \left[\overline{g^2} - \bar{g}\right]$$

and thus the normalized increase is

$$\frac{W_s}{\bar{x}} = \frac{1}{2} \bar{g} \left[\frac{\overline{g^2}}{(\bar{g})^2}\right] - \frac{1}{2} = \frac{1}{2} \bar{g}(1 + C_g^2) - \frac{1}{2}$$

[Note: a second method of proof for part (e) (using Little's result) is as follows: We first appeal to a known result to obtain the distribution of number in system for this bulk queue over all time. To this end define

$$P_k(t) = P[k \text{ in system at time } t]$$

and let

$$p_k = \lim_{t \to \infty} P_k(t) \quad \text{for} \quad k = 0, 1, 2, \ldots$$

Define the z-transform $P(z) \triangleq \sum_{k=0}^{\infty} p_k z^k$. Then, using the method of supplementary variables as in Exercise 5.7, it may be shown that

$$P(z) = (1-\rho) \frac{(1-z) B^*[\lambda - \lambda G(z)]}{B^*[\lambda - \lambda G(z)] - z}$$

5.12.

with
$$p_0 = 1 - \rho$$

(see [COHE 69] page 375, equation (2.13) for details). Note that $P(z) \neq Q(z)$ and $p_0 \neq d_0$. The average number in system $\bar{N}$ (over all time) is simply

$$\bar{N} = P^{(1)}(1) = \rho + \frac{\rho^2(1+C_b^2)}{2(1-\rho)} + \frac{\rho}{2(1-\rho)}[\bar{g}(1+C_g^2) - 1]$$

The average queue size $\bar{N}_q$ is

$$\bar{N}_q = \sum_{k=1}^{\infty}(k-1)p_k = \bar{N} - (1-p_0) = \bar{N} - \rho$$

or

$$\bar{N}_q = \frac{\rho^2(1+C_b^2)}{2(1-\rho)} + \frac{\rho}{2(1-\rho)}[\bar{g}(1+C_g^2) - 1]$$

We may now apply Little's result to $\bar{N}_q$ to find the average wait W. (Note that Little's result *cannot* be applied to $\bar{q}$, the average number in system at *departure* instants.) So $\bar{N}_q = \lambda W$ where $\lambda = \lambda \bar{g}$. Thus the average wait W is

$$W = \frac{\rho \bar{x}(1+C_b^2)}{2(1-\rho)} + \bar{x} \frac{\bar{g}(1+C_g^2) - 1}{2(1-\rho)}$$

and so

$$\frac{W}{\bar{x}} = \frac{\rho(1+C_b^2)}{2(1-\rho)} + \frac{\bar{g}(1+C_g^2) - 1}{2(1-\rho)}$$

We know, from part (d), that

$$\frac{W_g}{\bar{x}} = \frac{\rho \bar{g}}{2(1-\rho)}\left[1 + C_g^2 + \frac{C_b^2}{\bar{g}}\right]$$

Thus

$$\frac{W_s}{\bar{x}} = \frac{W - W_g}{\bar{x}} = \frac{\bar{g}(1+C_g^2) - 1}{2(1-\rho)} - \frac{\rho \bar{g}}{2(1-\rho)}(1+C_g^2) + \frac{\rho}{2(1-\rho)}$$

or

$$\frac{W_s}{\bar{x}} = \frac{1}{2}\bar{g}(1+C_g^2) - \frac{1}{2} \qquad \blacksquare$$

as before.]

5.12.

PROBLEM 5.13.

Consider an M/G/1 system in which service is instantaneous but is only available at "service instants", the intervals between successive service instants being independently distributed with PDF $F(x)$. The maximum number of customers that can be served at any service instant is m. Note that this is a bulk service system.

(a) Show that if q_n is the number of customers in the system just before the nth service instant, then

$$q_{n+1} = \begin{cases} q_n + v_n - m & q_n \geq m \\ v_n & q_n < m \end{cases}.$$

where v_n is the number of arrivals in the interval between the nth and $(n+1)$th service instants.

(b) Prove that the probability generating function of v_n is $F^*(\lambda - \lambda z)$. Hence show that $Q(z)$ is

$$Q(z) = \frac{\sum_{k=0}^{m-1} p_k (z^m - z^k)}{z^m [F^*(\lambda - \lambda z)]^{-1} - 1}$$

where $p_k = P[\tilde{q} = k]$ ($k = 0, \ldots, m-1$).

(c) The $\{p_k\}$ can be determined from the condition that within the unit disk of the z-plane, the numerator must vanish when the denominator does. Hence show that if $F(x) = 1 - e^{-\mu x}$,

$$Q(z) = \frac{z_m - 1}{z_m - z}$$

where z_m is the zero of $z^m[1 + \lambda(1-z)/\mu] - 1$ outside the unit disk.

SOLUTION

(a) If $q_n \geq m$, then m customers are served and clearly $q_{n+1} = q_n - m + v_n$. If $q_n < m$, everyone in the system gets served and $q_{n+1} = v_n$. Thus

$$q_{n+1} = \begin{cases} q_n + v_n - m & q_n \geq m \\ v_n & q_n < m \end{cases}$$

For convenience, we define the function $\Delta_{m,k}$ by

$$\Delta_{m,k} \triangleq \begin{cases} m & k \geq m \\ k & 0 \leq k \leq m \end{cases}$$

5.13.

Thus
$$q_{n+1} = q_n - \Delta_{m,q_n} + v_n.$$

(b) Since $F(x)$ is the distribution between successive service instants and $V(z)$ is the z-transform for the (Poisson) arrival process, the argument on page 197 may be applied. Thus $V(z) = F^*(\lambda - \lambda z)$. To find $Q(z)$, we follow the derivation in Section 5.6.

$$E\left[z^{q_{n+1}}\right] = E\left[z^{q_n - \Delta_{m,q_n} + v_n}\right] = E\left[z^{q_n - \Delta_{m,q_n}}\right] \cdot E\left[z^{v_n}\right]$$

since v_n, q_n are independent. Let $n \to \infty$ and obtain

$$Q(z) = E\left[z^{\tilde{q}}\right] = E\left[z^{\tilde{q} - \Delta_{m,\tilde{q}}}\right] \cdot E\left[z^{\tilde{v}}\right]$$

$$Q(z) = V(z) E\left[z^{\tilde{q} - \Delta_{m,\tilde{q}}}\right]$$

But

$$E\left[z^{\tilde{q} - \Delta_{m,\tilde{q}}}\right] = \sum_{k=0}^{m-1} P[\tilde{q} = k] z^{k-k} + \sum_{k=m}^{\infty} P[\tilde{q} = k] z^{k-m}$$

$$= \sum_{k=0}^{m-1} p_k + \frac{1}{z^m} \sum_{k=m}^{\infty} p_k z^k$$

$$= \sum_{k=0}^{m-1} p_k + \frac{1}{z^m} \left[Q(z) - \sum_{k=0}^{m-1} p_k z^k\right]$$

$$\therefore Q(z) = V(z) \left[\frac{1}{z^m} Q(z) + \sum_{k=0}^{m-1} p_k \left(1 - \frac{z^k}{z^m}\right)\right]$$

Recalling that $V(z) = F^*(\lambda - \lambda z)$ we have

$$Q(z) = F^*(\lambda - \lambda z) \frac{\sum_{k=0}^{m-1} p_k (z^m - z^k)}{z^m - F^*(\lambda - \lambda z)}$$

or

$$Q(z) = \frac{\sum_{k=0}^{m-1} p_k (z^m - z^k)}{z^m [F^*(\lambda - \lambda z)]^{-1} - 1}$$

5.13.

(c) Assume that $F(x) = 1 - e^{-\mu x}$. Therefore $F^*(\lambda - \lambda z) = \dfrac{\mu}{\lambda - \lambda z + \mu}$ and

$$Q(z) = \frac{\sum_{k=0}^{m-1} p_k(z^m - z^k)}{z^m \left[\dfrac{\lambda - \lambda z + \mu}{\mu}\right] - 1}$$

$$Q(z) = \frac{\sum_{k=0}^{m-1} p_k(z^m - z^k)}{z^m \left[1 + \dfrac{\lambda(1-z)}{\mu}\right] - 1} \triangleq \frac{P(z)}{D(z)}$$

We now examine the roots of the denominator

$$D(z) = z^m \left[1 + \frac{\lambda(1-z)}{\mu}\right] - 1.$$

Rewrite $D(z)$ as

$$D(z) = -\left[\frac{\lambda}{\mu} z^{m+1} - \left(1 + \frac{\lambda}{\mu}\right) z^m + 1\right]$$

$$= -[m\rho z^{m+1} - (1 + m\rho) z^m + 1]$$

where we write $\rho = \lambda/m\mu$. For stability we require $\lambda < m\mu$ or $\rho < 1$. We note that $D(z)$ is the negative of the denominator polynomial in Eq. (4.35) (with m replacing r), and thus we may use the results of Exercise 4.10. Thus $D(z)$ has one root at $z = 1$, $m-1$ roots in the range $|z| < 1$, and one root (say z_m) in $|z| > 1$. By analyticity of $Q(z)$ for $|z| \leqslant 1$, the m roots of $D(z)$ satisfying $|z| \leqslant 1$ cancel with the m numerator roots leaving

$$Q(z) = K \frac{1}{z_m - z}.$$

$Q(1) = 1$ implies $K = z_m - 1$. Thus

$$Q(z) = \frac{z_m - 1}{z_m - z}.$$

5.13.

PROBLEM 5.14.

Consider an M/G/1 system with bulk service. Whenever the server becomes free, he accepts *two* customers from the queue into service simultaneously, or, if only one is on queue, he accepts that one; in either case, the service time for the group (of size 1 or 2) is taken from $B(x)$. Let q_n be the number of customers remaining after the nth service instant. Let v_n be the number of arrivals during the nth service. Define $B^*(s)$, $Q(z)$, and $V(z)$ as transforms associated with the random variables $\tilde{x}$, $\tilde{q}$, and $\tilde{v}$ as usual. Let $\rho = \lambda \bar{x}/2$.

(a) Using the method of imbedded Markov chains, find
$$E(\tilde{q}) = \lim_{n \to \infty} E(q_n)$$
in terms of ρ, σ_b^2, and $P(\tilde{q}=0) \triangleq p_0$.

(b) Find $Q(z)$ in terms of $B^*(\cdot)$, p_0, and $p_1 \triangleq P(\tilde{q}=1)$.

(c) Express p_1 in terms of p_0.

SOLUTION

(a) Clearly we may write
$$q_{n+1} = \begin{cases} q_n - 2 + v_{n+1} & q_n \geqslant 2 \\ q_n - 1 + v_{n+1} & q_n = 1 \\ v_{n+1} & q_n = 0 \end{cases}$$

Introducing the function
$$\Delta_{2,k} \triangleq \begin{cases} 2 & k \geqslant 2 \\ k & 0 \leqslant k \leqslant 2 \end{cases}$$

we have $q_{n+1} = q_n - \Delta_{2,q_n} + v_{n+1}$. Letting $n \to \infty$ and taking expectations
$$\bar{q} = \bar{q} - E\left[\Delta_{2,\tilde{q}}\right] + \bar{v}$$

But
$$E\left[\Delta_{2,\tilde{q}}\right] = \sum_{k=0}^{\infty} \Delta_{2,k} P[\tilde{q}=k]$$
$$= P[\tilde{q}=1] + \sum_{k=2}^{\infty} 2\, P[\tilde{q}=k]$$
$$= p_1 + 2(1 - p_0 - p_1)$$

5.14.

So
$$\bar{v} = E\left[\Delta_{2,\tilde{q}}\right] = 2 - 2p_0 - p_1 \qquad \blacksquare$$

Recall:
$$q_{n+1} = q_n - \Delta_{2,q_n} + v_{n+1}$$

Square:
$$q_{n+1}^2 = q_n^2 - 2q_n\Delta_{2,q_n} + \Delta_{2,q_n}^2 + 2v_{n+1}(q_n - \Delta_{2,q_n}) + v_{n+1}^2$$

Let $n \to \infty$ and take expectations
$$\overline{q^2} = \overline{q^2} - 2E\left[\tilde{q}\Delta_{2,\tilde{q}}\right] + E\left[\Delta_{2,\tilde{q}}^2\right] + 2\bar{v}E\left[\tilde{q} - \Delta_{2,\tilde{q}}\right] + \overline{v^2}$$

since $\tilde{v}$ and $\tilde{q}$ are independent. So
$$2E\left[\tilde{q}\Delta_{2,\tilde{q}}\right] = E\left[\Delta_{2,\tilde{q}}^2\right] + \overline{v^2} + 2\bar{v}\left(\bar{q} - E\left[\Delta_{2,\tilde{q}}\right]\right)$$

Now $E\left[\Delta_{2,\tilde{q}}\right] = \bar{v}$ and
$$E\left[\Delta_{2,\tilde{q}}^2\right] = \sum_{k=1}^{\infty}\Delta_{2,k}^2 P[\tilde{q}=k]$$

$$= P[\tilde{q}=1] + 4\sum_{k=2}^{\infty}P[\tilde{q}=k]$$

$$= p_1 + 4(1 - p_0 - p_1)$$

Also
$$E\left[\tilde{q}\Delta_{2,\tilde{q}}\right] = \sum_{k=1}^{\infty} k\,\Delta_{2,k} P[\tilde{q}=k]$$

$$= P[\tilde{q}=1] + \sum_{k=2}^{\infty} 2k\,P[\tilde{q}=k]$$

$$= p_1 + 2\sum_{k=1}^{\infty} k\,P[\tilde{q}=k] - 2p_1$$

$$= 2\bar{q} - p_1$$

$$\therefore\ 2(2\bar{q} - p_1) = 4 - 4p_0 - 3p_1 + \overline{v^2} + 2\bar{v}(\bar{q} - \bar{v})$$

5.14.

Thus, using $p_1 = 2 - 2p_0 - \bar{v}$ we find
$$\bar{q} = \frac{2 - 2p_0 + \bar{v} + \overline{v^2} - 2(\bar{v})^2}{4 - 2\bar{v}}$$
But Eq. (5.46) gives $V(z) = B^*(\lambda - \lambda z)$. Thus
$$\bar{v} = V^{(1)}(1) = \lambda \bar{x} = 2\rho$$
by Eq. (5.58) and
$$\overline{v^2} - \bar{v} = V^{(2)}(1) = \lambda^2 \overline{x^2}$$
by Eq. (5.61).
$$\therefore \bar{q} = \frac{2(1 - p_0) + 2\rho + \lambda^2 \overline{x^2} + 2\rho - 2(4\rho^2)}{4 - 4\rho}$$

$$\bar{q} = \rho + \frac{2(1 - p_0) + \lambda^2 \overline{x^2} - 4\rho^2}{4(1 - \rho)}$$

$$\bar{q} = \rho + \frac{2(1 - p_0) + \lambda^2 \sigma_b^2}{4(1 - \rho)} \qquad \blacksquare$$

(b)
$$Q(z) = E\left[z^{\tilde{q}}\right] = E\left[z^{\tilde{q} - \Delta_{2,\tilde{q}} + \tilde{v}}\right]$$
$$= E\left[z^{\tilde{v}}\right] \cdot E\left[z^{\tilde{q} - \Delta_{2,\tilde{q}}}\right]$$
$$= V(z) E\left[z^{\tilde{q} - \Delta_{2,\tilde{q}}}\right]$$

But
$$E\left[z^{\tilde{q} - \Delta_{2,\tilde{q}}}\right] = \sum_{k=0}^{\infty} z^{k - \Delta_{2,k}} P[\tilde{q} = k]$$
$$= P[\tilde{q} = 0] + P[\tilde{q} = 1] + \sum_{k=2}^{\infty} z^{k-2} P[\tilde{q} = k]$$
$$= p_0 + p_1 + \frac{1}{z^2}[Q(z) - p_0 - p_1 z]$$

$$\therefore Q(z) = V(z)\left[p_0 + p_1 + \frac{1}{z^2}[Q(z) - p_0 - p_1 z]\right]$$

5.14.

$$Q(z) = V(z) \frac{p_0(1-z^2) + p_1 z(1-z)}{V(z) - z^2}$$

Therefore

$$Q(z) = B^*(\lambda - \lambda z) \frac{p_0(1-z^2) + p_1 z(1-z)}{B^*(\lambda - \lambda z) - z^2} \qquad \blacksquare$$

(c) From part (a),

$$\bar{v} = 2 - 2p_0 - p_1$$

But, from part (b) (using Eq. (5.58))

$$\bar{v} = \lambda \bar{x} = 2\rho$$

$$\therefore 2\rho = 2 - 2p_0 - p_1$$

$$p_1 = 2(1 - p_0 - \rho) \qquad \blacksquare$$

or

$$p_1 = 2(1 - p_0) - \lambda \bar{x} \qquad \blacksquare$$

PROBLEM 5.15.

Consider an M/G/1 queueing system with the following variation. The server refuses to serve any customers unless at least two customers are ready for service, at which time both are "taken into" service. These two customers are served individually and independently, one after the other. The instant at which the second of these two is finished is called a "critical" time and we shall use these critical times as the points in an imbedded Markov chain. Immediately following a critical time, if there are two more ready for service, they are both "taken into" service as above. If one or none are ready, then the server waits until a pair is ready, and so on. Let

q_n = number of customers left behind in the system immediately following the nth critical time

v_n = number of customers arriving during the combined service time of the nth *pair* of customers

(a) Derive a relationship between q_{n+1}, q_n, and v_{n+1}.

(b) Find
$$V(z) = \sum_{k=0}^{\infty} P[v_n = k] z^k$$

(c) Derive an expression for $Q(z) = \lim Q_n(z)$ as $n \to \infty$ in terms of $p_0 = P[\tilde{q} = 0]$, where
$$Q_n(z) = \sum_{k=0}^{\infty} P[q_n = k] z^k$$

(d) How would you solve for p_0?
(e) Describe (do *not* calculate) two methods for finding $\bar{q}$.

SOLUTION

(a) Since the server refuses to serve any customers unless at least two are in the system, we see that
$$q_{n+1} = \begin{cases} q_n - 2 + v_{n+1} & q_n \geqslant 2 \\ v_{n+1} & q_n \leqslant 1 \end{cases} \qquad \blacksquare$$

Introducing the function
$$\Delta_{2,k} \triangleq \begin{cases} 2 & k \geqslant 2 \\ k & 0 \leqslant k \leqslant 2 \end{cases}$$

we have
$$q_{n+1} = q_n - \Delta_{2, q_n} + v_{n+1}$$

(b) $V(z)$ is the z-transform for the number of arrivals in an interval which is the sum of two service times. The transform of the pdf for this interval is clearly $[B^*(s)]^2$. By analogy to the development of Figure 5.8 from Figure 5.7, we see that
$$V(z) = [B^*(\lambda - \lambda z)]^2 \qquad \blacksquare$$

(c) From part (a) we have,
$$Q_{n+1}(z) = E\left[z^{q_{n+1}}\right] = E\left[z^{q_n - \Delta_{2,q_n} + v_{n+1}}\right] = E\left[z^{v_{n+1}}\right] \cdot E\left[z^{q_n - \Delta_{2,q_n}}\right]$$
by independence of q_n and v_{n+1}. Letting $n \to \infty$, we have
$$Q(z) = E\left[z^{\tilde{q}}\right] = V(z) E\left[z^{\tilde{q} - \Delta_{2,\tilde{q}}}\right]$$

5.15.

But
$$E\left[z^{\tilde{q}-\Delta_{2,\tilde{q}}}\right] = \sum_{k=0}^{\infty} z^{k-\Delta_{2,k}} P[\tilde{q}=k]$$

$$= p_0 + p_1 + \sum_{k=2}^{\infty} z^{k-2} p_k$$

$$= p_0 + p_1 + \frac{1}{z^2}[Q(z) - p_0 - p_1 z]$$

Thus
$$Q(z) = \frac{V(z)}{z^2}\left[Q(z) + (z^2-1)p_0 + (z^2-z)p_1\right]$$

$$Q(z) = V(z)\frac{(1-z^2)p_0 + (z-z^2)p_1}{V(z)-z^2}$$

To eliminate p_1 we proceed as follows:
$$1 = \frac{Q(1)}{V(1)} = \lim_{z\to 1}\frac{(1-z^2)p_0 + (z-z^2)p_1}{V(z)-z^2}$$

Using L'Hospital's rule and $V^{(1)}(1) = \bar{v}$ we find
$$\bar{v} = 2 - 2p_0 - p_1$$

(This could also be obtained from the equation $\bar{q} = \bar{q} + E\left[\Delta_{2,\tilde{q}}\right] + \bar{v}$.) But $V(z) = [B^*(\lambda - \lambda z)]^2$ so that
$$\bar{v} = V^{(1)}(1) = 2\lambda\bar{x}.$$

$$\therefore\ 2\lambda\bar{x} = 2 - 2p_0 - p_1$$

and
$$p_1 = 2 - 2p_0 - 2\lambda\bar{x}.$$

Substituting this expression for p_1 into $Q(z)$ we have
$$Q(z) = [B^*(\lambda-\lambda z)]^2\frac{(1-z)[p_0(1-z) + 2z(1-\lambda\bar{x})]}{[B^*(\lambda-\lambda z)]^2 - z^2}\quad\blacksquare$$

(d) Equate roots of the denominator of $Q(z)$ with that of the numerator for $|z| < 1$ using analyticity of $Q(z)$.
(e) (1) $\bar{q} = Q^{(1)}(1)$
(2) Square the equation in part (a), let $n\to\infty$, and take expectations.

5.15.

PROBLEM 5.16.

Consider an M/G/1 queueing system in which service is given as follows. Upon entry into service, a coin is tossed, which has probability p of giving Heads. If the result is Heads, then the service time for that customer is zero seconds. If Tails, his service time is drawn from the following exponential distribution:

$$pe^{-px} \quad x \geq 0$$

(a) Find the average service time $\bar{x}$.
(b) Find the variance of service time σ_b^2.
(c) Find the expected waiting time W.
(d) Find $W^*(s)$.
(e) From (d), find the expected waiting time W.
(f) From (d), find $W(t) = P[\text{waiting time} \leq t]$.

SOLUTION

The service time density $b(x)$ is given by

$$b(x) = p\, u_0(x) + (1-p)\, p\, e^{-px} \quad x \geq 0$$

(a)
$$\bar{x} = 0 \cdot p + \frac{1}{p} \cdot (1-p) = \frac{1-p}{p} \qquad \blacksquare$$

(Thus, for stability, we require $p > \frac{\lambda}{\lambda+1}$.)

(b)
$$\overline{x^2} = 0 \cdot p + \frac{2}{p^2}(1-p) = \frac{2(1-p)}{p^2}$$

Thus

$$\sigma_b^2 = \overline{x^2} - (\bar{x})^2 = \frac{2(1-p)}{p^2} - \frac{(1-p)^2}{p^2}$$

$$\sigma_b^2 = \frac{1-p^2}{p^2} \qquad \blacksquare$$

(c) For M/G/1, $W = \dfrac{\lambda \overline{x^2}}{2(1-\rho)}$ by Eq. (5.70). Using $\rho = \lambda \bar{x} = \dfrac{\lambda(1-p)}{p}$ we have

5.16.

$$W = \frac{\lambda \dfrac{2(1-p)}{p^2}}{2\left[1 - \dfrac{\lambda(1-p)}{p}\right]}$$

$$W = \frac{\rho}{p(1-\rho)} \qquad \blacksquare$$

(d) We first note that

$$B^*(s) = p + (1-p)\frac{p}{s+p} = \frac{p(s+1)}{s+p}.$$

Thus, by Eq. (5.105),

$$W^*(s) = \frac{s(1-\rho)}{s - \lambda + \lambda \dfrac{p(s+1)}{s+p}}$$

$$W^*(s) = \frac{(1-\rho)(s+p)}{s + p(1-\rho)} \qquad \blacksquare$$

(e)

$$W^{*(1)}(s) = (1-\rho)\left[\frac{[s+p(1-\rho)] - (s+p)}{[s+p(1-\rho)]^2}\right]$$

$$= (1-\rho)\frac{-p\rho}{[s+p(1-\rho)]^2}$$

Thus

$$W = -W^{*(1)}(0) = \frac{\rho}{p(1-\rho)} \qquad \blacksquare$$

(same as part (c))

(f)

$$W^*(s) = \frac{(1-\rho)(s+p)}{s+p(1-\rho)}$$

$$= (1-\rho) + \frac{p\rho(1-\rho)}{s+p(1-\rho)}$$

Inverting we find the pdf as

$$w(y) = (1-\rho)u_0(y) + p\rho(1-\rho)e^{-p(1-\rho)y} \quad y \geq 0$$

5.16.

Thus the PDF is

$$W(y) = 1 - \rho e^{-\rho(1-\rho)y} \quad y \geq 0 \qquad \blacksquare$$

PROBLEM 5.17.†

Consider an M/G/1 queue. Let E be the event that T sec have elapsed since the arrival of the last customer. Begin at an arrival time and measure the time $\tilde{w}$ until event E next occurs. This measurement may involve the observation of many customer arrivals before E occurs.
(a) Let $\hat{A}(t)$ be the interarrival-time distribution for those intervals during which E does not occur. Find $\hat{A}(t)$.
(b) Find $\hat{A}^*(s) = \int_0^\infty e^{-st} d\hat{A}(t)$.
(c) Find $W^*(s|n) = \int_0^\infty e^{-sw} dW(w|n)$, where $W(w|n) = P[\text{time to event } E \leq w | n \text{ arrivals occur before } E]$.
(d) Find $W^*(s) = \int_0^\infty e^{-sw} dW(w)$, where $W(w) = P[\text{time to event } E \leq w]$.
(e) Find the mean time to event E.

SOLUTION

(a) If E does not occur, then the interarrival time must be $< T$ seconds. The probability that an interarrival time is $< T$ seconds is $1 - e^{-\lambda T}$. Thus

$$\hat{A}(t) = \begin{cases} \dfrac{1-e^{-\lambda t}}{1-e^{-\lambda T}} & 0 \leq t < T \\ 1 & t \geq T \end{cases} \qquad \blacksquare$$

$$\hat{a}(t) = \begin{cases} \dfrac{\lambda e^{-\lambda t}}{1-e^{-\lambda T}} & 0 \leq t < T \\ 0 & t \geq T \end{cases}$$

(b)

$$\hat{A}^*(s) = \int_0^\infty e^{-st} \hat{a}(t) \, dt = \int_0^T e^{-st} \frac{\lambda e^{-\lambda t}}{1 - e^{-\lambda T}} \, dt$$

$$\hat{A}^*(s) = \frac{\lambda}{s+\lambda} \cdot \frac{1 - e^{-(s+\lambda)T}}{1 - e^{-\lambda T}} \qquad \blacksquare$$

5.16. – 5.17.

(c) Recall that the random variable $\tilde{w}$ is defined to be the time to event E.
$$W(w|n) = P[\tilde{w} \leq w | n \text{ arrivals occur before } E]$$
Since exactly n arrivals occur before E, we know that the sum of these n interarrival times plus T must be the value of $\tilde{w}$. Thus
$$W(w|n) = P[n \text{ interarrival times} + T \leq w]$$
$$= P[t_1 + \cdots + t_n + T \leq w]$$
$$\therefore W^*(s|n) = E\left[e^{-s(t_1 + \cdots + t_n + T)}\right]$$
Noting that each interarrival time t_i occurs before E and thus has transform $\hat{A}^*(s)$, we have
$$W^*(s|n) = e^{-sT}[\hat{A}^*(s)]^n \qquad \blacksquare$$

(d)
$$W^*(s) = \sum_{n=0}^{\infty} W^*(s|n) \cdot P[n \text{ arrivals occur before } E]$$
$$= \sum_{n=0}^{\infty} e^{-sT}[\hat{A}^*(s)]^n [1-e^{-\lambda T}]^n e^{-\lambda T}$$
$$= e^{-(s+\lambda)T} \frac{1}{1 - \hat{A}^*(s)[1-e^{-\lambda T}]}$$
$$= e^{-(s+\lambda)T} \frac{1}{1 - \frac{\lambda}{s+\lambda}[1 - e^{-(s+\lambda)T}]}$$
$$W^*(s) = e^{-(s+\lambda)T} \frac{s+\lambda}{s + \lambda e^{-(s+\lambda)T}} \qquad \blacksquare$$

(e) We may calculate the mean time W to event E by $W = -W^{*(1)}(0)$. Instead let us find W as follows: from the argument on page 388, $W = \bar{N}\bar{t} + T$ where $\bar{t}$ is the mean interarrival time from the distribution $\hat{A}(t)$ and $\bar{N}$ is the mean number of such arrivals before E. So
$$\bar{N} = \sum_{k=1}^{\infty} k[1-e^{-\lambda T}]^k e^{-\lambda T} = e^{-\lambda T}[1-e^{-\lambda T}]\frac{1}{[e^{-\lambda T}]^2}$$
$$\bar{N} = e^{\lambda T}[1 - e^{-\lambda T}]$$

5.17.

$$\bar{t} = \int_0^\infty [1 - \hat{A}(t)]\, dt = \int_0^T \frac{e^{-\lambda t} - e^{-\lambda T}}{1 - e^{-\lambda T}}\, dt$$

$$\bar{t} = \frac{1}{1 - e^{-\lambda T}} \left[\frac{1 - e^{-\lambda T}}{\lambda} - T e^{-\lambda T} \right]$$

Thus

$$W = e^{\lambda T}[1 - e^{-\lambda T}] \frac{1}{1 - e^{-\lambda T}} \left[\frac{1 - e^{-\lambda T}}{\lambda} - T e^{-\lambda T} \right] + T = \frac{e^{\lambda T} - 1}{\lambda} - T + T$$

$$W = \frac{e^{\lambda T} - 1}{\lambda} \qquad \blacksquare$$

PROBLEM 5.18.

Consider an M/G/1 system in which time is divided into intervals of length q sec each. Assume that *arrivals* are Bernoulli, that is,

$$P[1 \text{ arrival in any interval}] = \lambda q$$
$$P[0 \text{ arrivals in any interval}] = 1 - \lambda q$$
$$P[> 1 \text{ arrival in any interval}] = 0$$

Assume that a customer's *service time* $\tilde{x}$ is some multiple of q sec such that

$$P[\text{service time} = nq \text{ sec}] = g_n \quad n = 0, 1, 2, \ldots$$

(a) Find $E[\text{number of arrivals in an interval}]$.
(b) Find the average arrival rate.
(c) Express $E[\tilde{x}] \triangleq \bar{x}$ and $E[\tilde{x}(\tilde{x} - q)] \triangleq \overline{x^2} - \bar{x}q$ in terms of the moments of the g_n distribution (i.e., let $\overline{g^k} \triangleq \sum_{n=0}^\infty n^k g_n$).
(d) Find $y_{mn} = P[m \text{ customers arrive in } nq \text{ sec}]$.
(e) Let $v_m = P[m \text{ customers arrive during the service of a customer}]$ and let

$$V(z) = \sum_{m=0}^\infty v_m z^m \quad \text{and} \quad G(z) = \sum_{m=0}^\infty g_m z^m$$

Express $V(z)$ in terms of $G(z)$ and the system parameters λ and q.
(f) Find the mean number of arrivals during a customer service time from (e).

5.17. – 5.18.

SOLUTION

(a)
$$E[\text{number of arrivals in an interval}] = 1 \cdot \lambda q + 0 \cdot (1 - \lambda q) + 0 = \lambda q \quad \blacksquare$$

(b)
$$\text{average arrival rate} = \frac{\lambda q}{q} = \lambda \quad \blacksquare$$

(c)
$$\bar{x} = \sum_{n=0}^{\infty} (nq) g_n = q\,\bar{g}$$

$$\overline{x^2} = \sum_{n=0}^{\infty} (nq)^2 g_n = q^2 \overline{g^2}$$

$$\therefore \overline{x^2} - \bar{x}\,q = q^2 (\overline{g^2} - \bar{g}) \quad \blacksquare$$

(d) In nq seconds there are n intervals. We want m intervals each with one arrival and $n - m$ with none. Thus

$$y_{mn} = \binom{n}{m} (\lambda q)^m (1 - \lambda q)^{n-m} \quad m \leq n \quad \blacksquare$$

$$(y_{mn} = 0 \quad \text{for} \quad m > n)$$

(e) Condition on the service time $\tilde{x}$. Then

$$v_m = \sum_{n=0}^{\infty} g_n\, P[v_m | \tilde{x} = nq] = \sum_{n=0}^{\infty} g_n\, y_{mn}$$

and thus

$$V(z) = \sum_{m=0}^{\infty} v_m z^m = \sum_{m=0}^{\infty} \sum_{n=0}^{\infty} g_n y_{mn} z^m$$

$$= \sum_{m=0}^{\infty} \sum_{n=m}^{\infty} g_n z^m \binom{n}{m} (\lambda q)^m (1 - \lambda q)^{n-m}$$

$$= \sum_{n=0}^{\infty} g_n \sum_{m=0}^{n} \binom{n}{m} (\lambda q z)^m (1 - \lambda q)^{n-m}$$

$$= \sum_{n=0}^{\infty} g_n (\lambda q z + 1 - \lambda q)^n$$

5.18.

$$V(z) = G[1 - \lambda q(1-z)]$$

(f)

$$\bar{v} = V^{(1)}(1) = G^{(1)}[1 - \lambda q(1-z)](\lambda q)\Big|_{z=1}$$

$$= G^{(1)}(1)(\lambda q)$$

$$\bar{v} = \lambda q \bar{g}$$

i.e. $\bar{v} = \lambda \bar{x} = \rho$

PROBLEM 5.19.

Suppose that in an M/G/1 queueing system the *cost* of making a customer wait t sec is $c(t)$ dollars, where $c(t) = \alpha e^{\beta t}$. Find the average cost of queueing for a customer. Also determine the conditions necessary to keep the average cost finite.

SOLUTION

Clearly β is constrained to be real. The average cost of queueing is simply

$$\bar{c} = \int_0^\infty c(t)\, dW(t)$$

$$\bar{c} = \alpha \int_0^\infty e^{\beta t}\, dW(t).$$

Assuming this integral exists, $\bar{c} = \alpha W^*(-\beta)$. For M/G/1, we may use the P–K transform equation (Eq. (5.105)) to obtain

$$\bar{c} = \alpha \frac{(-\beta)(1-\rho)}{-\beta - \lambda + \lambda B^*(-\beta)}$$

or

$$\bar{c} = \frac{\alpha \beta (1-\rho)}{\beta + \lambda - \lambda B^*(-\beta)}$$

We must next determine conditions for the existence of the integral

$$I \triangleq \int_0^\infty e^{\beta t}\, dW(t)$$

From the discussion in Section I.3, we see that there is a value β_0 such that $W^*(-\beta) < \infty$ for $\beta < \beta_0$. Since I clearly exists for $\beta \leq 0$, we need merely study $0 < \beta < \beta_0$. If $I < \infty$, then for M/G/1 we have

$$I = W^*(-\beta) = \frac{\beta(1-\rho)}{\beta + \lambda - \lambda B^*(-\beta)}.$$

In this case, β_0 must then be the smallest positive pole of $W^*(-\beta)$. That is, β_0 is the smallest positive root of the equation $\beta + \lambda - \lambda B^*(-\beta) = 0$, and so it must be that

$$\beta_0 = \lambda[B^*(-\beta_0) - 1].$$

[Note: Let us give a rigorous argument for the existence of such a β_0. Define

$$f(\beta) \triangleq \lambda[B^*(-\beta) - 1]$$

and note (by differentiation) that $f(\beta)$ is convex increasing in β, $f(0) = 0$, and $f'(0) = \rho < 1$ for a stable system. Also,

$$f'(\beta) = \lambda \left(\int_0^\infty x\, e^{\beta x} b(x)\, dx \right)$$

$$= \lambda \left(\int_0^\infty x \left[\sum_{k=0}^\infty \frac{(\beta x)^k}{k!} \right] b(x)\, dx \right)$$

$$= \lambda \sum_{k=0}^\infty \frac{\overline{x^{k+1}}}{k!} \beta^k$$

Thus, for $\beta \geq 0$,

$$f'(\beta) \geq \lambda \left[\bar{x} + \beta \overline{x^2} \right] \geq \lambda \left[\bar{x} + \beta (\bar{x})^2 \right]$$

So, for $\lambda > 0$ and $\bar{x} > 0$ (i.e. $\rho > 0$), we have

$$\lim_{\beta \to \infty} f'(\beta) = \infty.$$

Thus, as shown in the following figure, the equation $\beta = f(\beta)$ has a solution $\beta_0 > 0$ for $0 < \rho < 1$.]

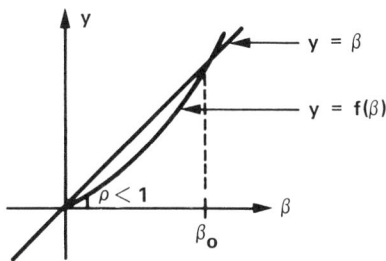

5.19.

In summary, for M/G/1 $(0 < \rho < 1)$

$$\bar{c} = \frac{\alpha \beta (1-\rho)}{\beta + \lambda - \lambda B^*(-\beta)}$$

and $\bar{c}$ is finite for $\beta < \beta_0$ where β_0 is the smallest positive root of the equation $\beta + \lambda - \lambda B^*(-\beta) = 0$.

PROBLEM 5.20.

We wish to find the *interdeparture* time probability density function $d(t)$ for an M/G/1 queueing system.
(a) Find the Laplace transform $D^*(s)$ of this density conditioned first on a nonempty queue left behind, and second on an empty queue left behind by a departing customer. Combine these results to get the Laplace transform of the interdeparture time density and from this find the density itself.
(b) Give an explicit form for the probability distribution $D(t)$, or density $d(t) = dD(t)/dt$, of the interdeparture time when we have a constant service time, that is

$$B(x) = \begin{cases} 0 & x < T \\ 1 & x \geq T \end{cases}$$

SOLUTION

(a) Proceeding as on page 148 in Section 4.8 we have

$$D^*(s)\Big|_{\substack{\text{non-empty queue} \\ \text{left behind}}} = B^*(s)$$

$$D^*(s)\Big|_{\substack{\text{empty queue} \\ \text{left behind}}} = B^*(s) A^*(s) = B^*(s) \frac{\lambda}{s+\lambda}$$

Thus, unconditioning, we have

$$D^*(s) = \rho B^*(s) + (1-\rho) B^*(s) \frac{\lambda}{s+\lambda} \qquad \blacksquare$$

or

$$D^*(s) = \frac{\rho s + \lambda}{s + \lambda} B^*(s) \qquad \blacksquare$$

From the first of these two forms we invert by inspection to obtain the interdeparture time density as

$$d(t) = \rho\, b(t) + (1-\rho)\, b(t) \circledast \lambda e^{-\lambda t} \quad t \geq 0 \qquad \blacksquare$$

(b) Since we have a constant service time of T seconds, we have

$$b(t) = u_0(t - T) \iff e^{-sT} = B^*(s).$$

We may proceed from the expression for $D^*(s)$ or from $d(t)$ in part (a). Let us use the latter approach:

$$\begin{aligned}
b(t) \circledast \lambda e^{-\lambda t} &= \int_0^t b(t-x)\, \lambda e^{-\lambda x}\, dx \\
&= \int_0^t u_0(t - x - T)\, \lambda e^{-\lambda x}\, dx \\
&= \lambda e^{-\lambda(t-T)} \delta(t-T)
\end{aligned}$$

So

$$d(t) = \rho\, u_0(t-T) + (1-\rho)\, \lambda e^{-\lambda(t-T)} \delta(t-T) \qquad \blacksquare$$

and thus

$$D(t) = \begin{cases} 0 & t < T \\ 1 - (1-\rho) e^{-\lambda(t-T)} & t \geq T \end{cases} \qquad \blacksquare$$

PROBLEM 5.21.

Consider the following modified order of service for M/G/1. Instead of LCFS as in Figure 5.11, assume that after the interval x_1, the sub-busy period generated by C_2 occurs, which is followed by the sub-busy period generated by C_3, and so on, until the busy period terminates. Using the sequence of arrivals and service times shown in the upper contour of Figure 5.11a, redraw parts a, b, and c to correspond to the above order of service.

SOLUTION

Careful examination of Figure 5.11 reveals the arrival times (τ_i) and service times (x_i) listed in the table below. (Here we have assumed that $\tau_1 = 1$.) Let us study the algorithm under consideration to obtain the departure times. First, note that the starter of a busy period is served completely (C_1 in our case). All customers who arrive during his service time are set aside (C_2, C_3 and C_4).

Each of these then generates his own sub-busy period in order (where service is assumed to be FCFS within each sub-busy period). This analysis yields the following chart:

Customer	1	2	3	4	5	6	7	8	9	10	11
Arrival Time	1	3	4	5	7	8	13	15	17	23	27
Service Time	5	6	2	3	2	1	2	2	2	3	1
Departure Time	6	12	23	29	14	15	17	19	21	26	30

Note: We have resolved the unfortunate "tie" at $t=23$ by anticipating the convention introduced later in Section 8.4 (on page 304), namely that an idle period must have non-zero length. Thus we consider C_{10} to be part of C_3's sub-busy period. If we assume zero-length idle periods, then C_{10} would be part of C_4's sub-busy period and would be served after C_4 and not before as we have assumed in our solution.

(a) Decomposition of the busy period

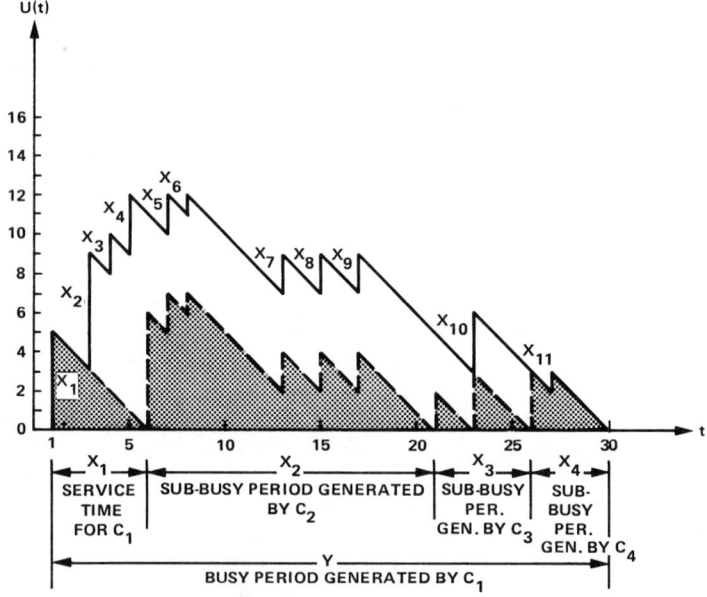

5.21.

(b) Number in the system

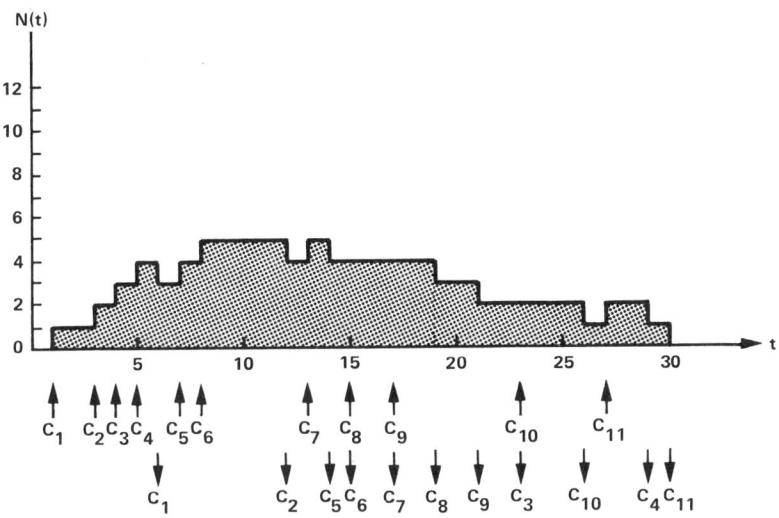

(c) Customer history

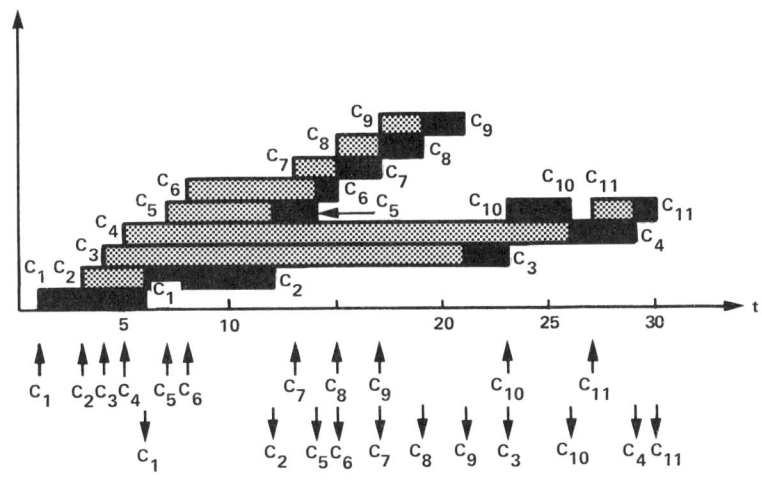

5.21.

PROBLEM 5.22.

Consider an M/G/1 system in which a departing customer immediately joins the queue again with probability p, or departs forever with probability $q = 1 - p$. Service is FCFS, and the service time for a returning customer is independent of his previous service times. Let $B^*(s)$ be the transform for the service time pdf and let $B_T^*(s)$ be the transform for a customer's *total* service time pdf.

(a) Find $B_T^*(s)$ in terms of $B^*(s)$, p, and q.
(b) Let $\overline{x_T^n}$ be the nth moment of the total service time. Find $\overline{x_T^1}$ and $\overline{x_T^2}$ in terms of $\bar{x}$, $\overline{x^2}$, p, and q.
(c) Show that the following recurrence formula holds:

$$\overline{x_T^n} = \overline{x^n} + \frac{p}{q} \sum_{k=1}^{n} \binom{n}{k} \overline{x^k}\, \overline{x_T^{n-k}}$$

(d) Let

$$Q_T(z) = \sum_{k=0}^{\infty} p_{kT} z^k$$

where $p_{kT} = P[\text{number in system} = k]$. For $\lambda \bar{x} < q$ prove that

$$Q_T(z) = \left(1 - \frac{\lambda \bar{x}}{q}\right) \frac{q(1-z) B^*[\lambda(1-z)]}{(q+pz) B^*[\lambda(1-z)] - z}$$

(e) Find $\bar{N}$, the average number of customers in the system.

SOLUTION

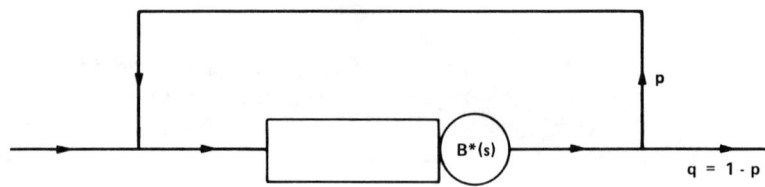

(a) Conditioning on the number of return trips, we have

$$B_T^*(s \mid \text{exactly } n \text{ return trips}) = [B^*(s)]^{n+1}$$

5.22.

Unconditioning yields

$$B_T^*(s) = \sum_{n=0}^{\infty} B_T^*(s|\text{exactly } n \text{ return trips}) \cdot P(\text{exactly } n \text{ return trips})$$

$$= \sum_{n=0}^{\infty} [B^*(s)]^{n+1} q\, p^n = q\, B^*(s) \sum_{n=0}^{\infty} [p\, B^*(s)]^n$$

$$B_T^*(s) = \frac{q\, B^*(s)}{1 - p\, B^*(s)} \qquad \blacksquare$$

(b) Differentiating, we obtain

$$B_T^{*(1)}(s) = \frac{[1 - p\, B^*(s)]\, q\, B^{*(1)}(s) - q\, B^*(s)[-p\, B^{*(1)}(s)]}{[1 - p\, B^*(s)]^2}$$

$$= \frac{q\, B^{*(1)}(s)}{[1 - p\, B^*(s)]^2}$$

Thus

$$\overline{x_T^1} = -B_T^{*(1)}(0) = -\frac{q\,(-\bar{x})}{(1-p)^2}$$

$$\overline{x_T^1} = \frac{\bar{x}}{q} \qquad \blacksquare$$

The second derivative gives

$$B_T^{*(2)}(s) = \frac{[1-p\,B^*(s)]^2 q\, B^{*(2)}(s) - q\, B^{*(1)}(s)\, 2\,[1-p\,B^*(s)]\,[-p\,B^{*(1)}(s)]}{[1-p\,B^*(s)]^4}$$

$$= \frac{[1-p\,B^*(s)]\, q\, B^{*(2)}(s) + 2pq\, [B^{*(1)}(s)]^2}{[1-p\,B^*(s)]^3}$$

Thus

$$\overline{x_T^2} = B_T^{*(2)}(0) = \frac{(1-p)\, q\, \overline{x^2} + 2pq\,(-\bar{x})^2}{(1-p)^3}$$

$$\overline{x_T^2} = \frac{\overline{x^2}}{q} + \frac{2p(\bar{x})^2}{q^2} \qquad \blacksquare$$

(c) From

$$B_T^*(s) = \frac{q\, B^*(s)}{1 - p\, B^*(s)}$$

5.22.

we have
$$B_T^*(s) = q B^*(s) + p B^*(s) B_T^*(s)$$

Thus
$$B_T^{*(n)}(s) = q B^{*(n)}(s) + p [B^*(s) B_T^*(s)]^{(n)}$$

Using the formula for the nth derivative of the product of two functions, namely,
$$(fg)^{(n)} = \sum_{k=0}^{n} \binom{n}{k} f^{(k)} g^{(n-k)}$$

we have
$$B_T^{*(n)}(s) = q B^{*(n)}(s) + p \sum_{k=0}^{n} \binom{n}{k} B^{*(k)}(s) B_T^{*(n-k)}(s)$$

Setting $s=0$ we obtain
$$B_T^{*(n)}(0) = q B^{*(n)}(0) + p \sum_{k=0}^{n} \binom{n}{k} B^{*(k)}(0) B_T^{*(n-k)}(0)$$

or
$$(-1)^n \overline{x_T^n} = q(-1)^n \overline{x^n} + p \sum_{k=0}^{n} \binom{n}{k} (-1)^k \overline{x^k} \cdot (-1)^{n-k} \overline{x_T^{n-k}}$$

Thus
$$\overline{x_T^n} = q \overline{x^n} + p \sum_{k=0}^{n} \binom{n}{k} \overline{x^k}\, \overline{x_T^{n-k}}$$

or
$$\overline{x_T^n} = q \overline{x^n} + p \overline{x_T^n} + p \sum_{k=1}^{n} \binom{n}{k} \overline{x^k}\, \overline{x_T^{n-k}}$$

$$\therefore\ \overline{x_T^n} = \overline{x^n} + \frac{p}{q} \sum_{k=1}^{n} \binom{n}{k} \overline{x^k}\, \overline{x_T^{n-k}}$$

(d) In determining the number in the system, we may assume (with impunity) that a customer cycles back directly into service instead of to the tail of the queue. This is allowed due to the "memoryless" selection of a new service time each time a customer returns in addition to the independence of the feedback decision. Thus we may consider our queue as an M/G/1 system with $B_T^*(s)$ as the transform for service time, and so Eq. (5.86) may be applied to determine $Q_T(z)$. (Note: This will not work for the determination of $W_T^*(s)$.) Thus

5.22.

$$Q_T(z) = B_T^*(\lambda - \lambda z) \frac{(1-\rho_T)(1-z)}{B_T^*(\lambda - \lambda z) - z}$$

where

$$B_T^*(s) = \frac{q\, B^*(s)}{1 - p\, B^*(s)}$$

and

$$\rho_T = \lambda\, \overline{x_T^1} = \frac{\lambda\, \overline{x}}{q}$$

So

$$Q_T(z) = \frac{q\, B^*(\lambda - \lambda z)}{1 - p\, B^*(\lambda - \lambda z)} \cdot \frac{\left(1 - \frac{\lambda\, \overline{x}}{q}\right)(1-z)}{\frac{q\, B^*(\lambda - \lambda z)}{1 - p\, B^*(\lambda - \lambda z)} - z}$$

$$Q_T(z) = \left(1 - \frac{\lambda\, \overline{x}}{q}\right) \frac{q(1-z)\, B^*(\lambda - \lambda z)}{(q + pz)\, B^*(\lambda - \lambda z) - z}$$

(e) For M/G/1, $\overline{N} = \overline{q} = \rho + \dfrac{\lambda^2\, \overline{x^2}}{2(1-\rho)}$. Thus

$$\overline{N} = \rho_T + \frac{\lambda^2\, \overline{x_T^2}}{2(1-\rho_T)}$$

From our earlier results, we have

$$\overline{N} = \frac{\lambda\, \overline{x}}{q} + \frac{\lambda^2 \left[\dfrac{\overline{x^2}}{q} + 2p\, \dfrac{(\overline{x})^2}{q^2}\right]}{2\left(1 - \dfrac{\lambda\, \overline{x}}{q}\right)}$$

$$= \frac{2\lambda\, \overline{x}(q - \lambda\, \overline{x}) + \lambda^2 [q\, \overline{x^2} + 2p(\overline{x})^2]}{2q\, (q - \lambda\, \overline{x})}$$

$$= \frac{2\lambda\, \overline{x}\, q + \lambda^2 q\, \overline{x^2} - 2\lambda^2 (\overline{x})^2 (1 - p)}{2q\, (q - \lambda\, \overline{x})}$$

5.22.

Therefore

$$\overline{N} = \frac{2\lambda \overline{x}(1-\lambda \overline{x}) + \lambda^2 \overline{x^2}}{2(q - \lambda \overline{x})}$$ ∎

PROBLEM 5.23.†

Consider a first-come-first-served M/G/1 queue with the following changes. The server serves the queue as long as someone is in the system. Whenever the system empties the server goes away on vacation for a certain length of time, which may be a random variable. At the end of his vacation the server returns and begins to serve customers again; if he returns to an empty system then he goes away on vacation again. Let $F(z) = \sum_{j=1}^{\infty} f_j z^j$ be the z-transform for $\tilde{f}$, the number of customers awaiting service when the server returns from vacation to find at least one customer waiting (that is, f_j is the probability that at the initiation of a busy period the server finds j customers awaiting service).

(a) Derive an expression which gives q_{n+1} in terms of q_n, v_{n+1}, and $\tilde{f}$ (the number of customer arrivals during the server's vacation).

(b) Derive an expression for $Q(z)$ where $Q(z) = \lim E[z^{q_n}]$ as $n \to \infty$ in terms of p_0 (equal to the probability that a departing customer leaves 0 customers behind).

(c) Show that $p_0 = (1-\rho)/F^{(1)}(1)$ where $F^{(1)}(1) = \partial F(z)/\partial z |_{z=1}$ and $\rho = \lambda \overline{x}$.

(d) Assume now that the service vacation will end whenever a new customer enters the empty system. For this case find $F(z)$ and show that when we substitute it back into our answer for (b) then we arrive at the classical M/G/1 solution.

SOLUTION

(a) Clearly, if $q_n > 0$, then $q_{n+1} = q_n - 1 + v_{n+1}$ as for the usual M/G/1 system. If $q_n = 0$ the server goes on vacation and does not again begin serving until there are $\tilde{f} \geq 1$ in the system (recall that $f_j \Leftrightarrow F(z)$). Thus $q_{n+1} = \tilde{f} - 1 + v_{n+1}$ for $q_n = 0$. Define a random variable

$$\Delta_{\tilde{f}, k} \triangleq \begin{cases} k & k > 0 \\ \tilde{f} & k = 0 \end{cases}$$

Then
$$q_{n+1} = v_{n+1} - 1 + \Delta_{\tilde{f}, q_n}.$$ ∎

(b)
$$Q_{n+1}(z) = E\left[z^{q_{n+1}}\right] = E\left[z^{v_{n+1} - 1 + \Delta_{\tilde{f}, q_n}}\right]$$
$$= E\left[z^{v_{n+1} - 1}\right] \cdot E\left[z^{\Delta_{\tilde{f}, q_n}}\right]$$

by independence of v_{n+1}, $\tilde{f}$, and q_n. Letting $n \to \infty$ we find
$$Q(z) = \frac{V(z)}{z} E\left[z^{\Delta_{\tilde{f}, \tilde{q}}}\right]$$

But
$$E\left[z^{\Delta_{\tilde{f}, \tilde{q}}}\right] = E[z^{\tilde{f}}] \cdot P[\tilde{q} = 0] + \sum_{k=1}^{\infty} z^k P[\tilde{q} = k]$$
$$= F(z) p_0 + [Q(z) - p_0]$$

Thus
$$Q(z) = \frac{V(z)}{z}\left[[F(z) - 1]p_0 + Q(z)\right]$$

or
$$Q(z) = V(z) \frac{p_0[1 - F(z)]}{V(z) - z}$$ ∎

[Note that this equation, and its derivation, is the same as that used in the proof of Exercise 5.12(b). Here, $V(z) = B^*(\lambda - \lambda z)$ whereas in 5.12(b), $V(z) = B^*[\lambda - \lambda G(z)]$. Thus, for $Q(z)$, this vacation problem is equivalent to an M/G/1 bulk arrival system where bulks may only arrive at the initiation of a busy period.] Finally,

$$Q(z) = B^*(\lambda - \lambda z) \frac{p_0[1 - F(z)]}{B^*(\lambda - \lambda z)}$$ ∎

(c) From part (b) we have
$$\frac{Q(z)}{V(z)} = \frac{p_0[1 - F(z)]}{V(z) - z}$$

To determine p_0 we evaluate the above equation at $z = 1$ (using L'Hospital's rule). Then

5.23.

$$1 = \frac{Q(1)}{V(1)} = \lim_{z \to 1} \frac{p_0[1 - F(z)]}{B^*(\lambda - \lambda z) - z}$$

$$= \frac{p_0[-F^{(1)}(1)]}{B^{*(1)}(0)(-\lambda) - 1} = \frac{p_0 F^{(1)}(1)}{1 - \lambda \bar{x}}$$

So

$$p_0 = \frac{1 - \rho}{F^{(1)}(1)}$$

where $\rho = \lambda \bar{x}$.

(d) In this case, $f_1 = 1$ and $f_k = 0$ for $k > 1$. Thus

$$F(z) = z \qquad \blacksquare$$

Hence $F^{(1)}(1) = 1$ and $p_0 = 1 - \rho$. So

$$Q(z) = B^*(\lambda - \lambda z) \frac{(1 - \rho)(1 - z)}{B^*(\lambda - \lambda z) - z} \qquad \blacksquare$$

which is the P–K transform equation for M/G/1.

PROBLEM 5.24.

We recognize that an arriving customer who finds k others in the system is delayed by the remaining service time for the customer in service plus the sum of $(k-1)$ complete service times.

(a) Using the notation and approach of Exercise 5.7, show that we may express the transform of the waiting time pdf as

$$W^*(s) = p_0 + \int_0^\infty \sum_{k=1}^\infty p_k(x_0)[B^*(s)]^{k-1}$$

$$\times \int_0^\infty e^{-sy} r(y + x_0) e^{-\int_0^{y+x_0} r(u)\, du}\, dy$$

$$\times e^{\int_0^{x_0} r(u)\, du}\, dx_0$$

(b) Show that the expression in (a) reduces to $W^*(s)$ as given in Eq. (5.106).

5.23. – 5.24.

SOLUTION

(a) $W^*(s) = E[e^{-s\tilde{w}}]$ where $\tilde{w}$ is the waiting time of a customer. Since $P[\tilde{w}=0] = p_0$ (Poisson arrivals) we have $W^*(s) = p_0 + E[e^{-s\tilde{w}}|\tilde{w}>0]$. Assuming a customer has to wait, his waiting time is the sum of the residual service time of the customer being served plus the service times of all those in queue. Thus, if he arrives to find k in the system, his wait is $\tilde{x}_1 + \cdots + \tilde{x}_{k-1} + \tilde{r}$ where the $\tilde{x}_i$ have pdf $b(x)$ with transform $B^*(s)$ and $\tilde{r}$ is the residual life of the customer in service. Given that this customer has received x_0 seconds already, we find

$$P[y < \tilde{r} \leq y + \Delta y | x_0] = \frac{b(y+x_0)}{1-B(x_0)} \Delta y$$

So

$$E[e^{-s\tilde{w}}|k, x_0, \tilde{w}>0] = E[e^{-s(\tilde{x}_1 + \cdots + \tilde{x}_{k-1} + \tilde{r})}|x_0, \tilde{w}>0]$$

$$= E[e^{-s\tilde{x}_1}] \cdots E[e^{-s\tilde{x}_{k-1}}] \cdot E[e^{-s\tilde{r}}|x_0, \tilde{w}>0]$$

$$= [B^*(s)]^{k-1} \times \int_0^\infty e^{-sy} \frac{b(y+x_0)}{1-B(x_0)} dy$$

Unconditioning on k and x_0 gives

$$E[e^{-s\tilde{w}}|\tilde{w}>0] = \int_0^\infty \sum_{k=1}^\infty p_k(x_0)[B^*(s)]^{k-1} \times \int_0^\infty e^{-sy} \frac{b(y+x_0)}{1-B(x_0)} dy\, dx_0$$

Recalling from Exercise 5.7(e) that

$$b(y+x_0) = r(y+x_0) e^{-\int_0^{y+x_0} r(u)\, du}$$

$$1 - B(x_0) = e^{-\int_0^{x_0} r(u)\, du}$$

we have the desired equation

$$W^*(s) = p_0 + \int_0^\infty \sum_{k=1}^\infty p_k(x_0)[B^*(s)]^{k-1}$$

$$\times \int_0^\infty e^{-sy} r(y+x_0) e^{-\int_0^{y+x_0} r(u)\, du} dy$$

$$\times e^{\int_0^{x_0} r(u)\, du} dx_0$$

5.24.

(b) We recall from part (a) that

$$W^*(s) = p_0 + \int_0^\infty \sum_{k=1}^\infty p_k(x_0)[B^*(s)]^{k-1}$$

$$\times \int_0^\infty e^{-sy} b(y+x_0)\, dy \, \frac{1}{1-B(x_0)}\, dx_0$$

To simplify further calculations we define a function

$$F(z) = \int_0^\infty \sum_{k=1}^\infty p_k(x_0) z^k \int_0^\infty e^{-sy} b(y+x_0)\, dy \, \frac{1}{1-B(x_0)}\, dx_0$$

Note that

$$W^*(s) = p_0 + \frac{F(B^*(s))}{B^*(s)}$$

and so we must determine $F(z)$. Recalling from Exercise 5.7(d) that

$$R(z, x_0) \triangleq \sum_{k=1}^\infty p_k(x_0) z^k$$

we have

$$F(z) = \int_0^\infty R(z, x_0) \int_0^\infty e^{-sy} b(y+x_0)\, dy \, \frac{1}{1-B(x_0)}\, dx_0$$

By Exercise 5.7(e), we have

$$R(z, x_0) = R(z, 0)\, e^{-\lambda x_0 (1-z) - \int_0^{x_0} r(y)\, dy}$$

$$= R(z, 0)\, e^{-\lambda x_0 (1-z)} [1 - B(x_0)]$$

Thus

$$F(z) = R(z, 0) \int_0^\infty e^{-\lambda x_0 (1-z)} \int_0^\infty e^{-sy} b(y+x_0)\, dy\, dx_0$$

$$= R(z, 0) \int_0^\infty e^{-\lambda x_0 (1-z)} \int_{x_0}^\infty e^{-s(t-x_0)} b(t)\, dt\, dx_0$$

Interchanging the order of integration gives

$$F(z) = R(z, 0) \int_0^\infty e^{-st} b(t) \int_0^t e^{-x_0(\lambda - \lambda z - s)}\, dx_0\, dt$$

Carrying out the innner integration yields

$$F(z) = R(z, 0) \int_0^\infty e^{-st} b(t) \left[\frac{e^{-t(\lambda - \lambda z - s)} - 1}{s - \lambda(1-z)} \right] dt$$

5.24.

and so

$$F(z) = R(z,0) \frac{\left[\int_0^\infty e^{-t(\lambda - \lambda z)} b(t) \, dt - \int_0^\infty e^{-st} b(t) \, dt\right]}{s - \lambda(1-z)}$$

or

$$F(z) = R(z,0) \frac{B^*(\lambda - \lambda z) - B^*(s)}{s - \lambda(1-z)}$$

Exercise 5.7(e) gives

$$R(z,0) = \frac{\lambda z(z-1) p_0}{z - B^*(\lambda - \lambda z)}$$

Thus

$$F(z) = p_0 \frac{\lambda z(z-1)}{s - \lambda(1-z)} \cdot \frac{B^*(\lambda - \lambda z) - B^*(s)}{z - B^*(\lambda - \lambda z)}$$

Evaluating F at the point $B^*(s)$ gives

$$F(B^*(s)) = p_0 \frac{\lambda B^*(s)[1 - B^*(s)]}{s - \lambda[1 - B^*(s)]}$$

Using

$$W^*(s) = p_0 + \frac{F(B^*(s))}{B^*(s)}$$

we find

$$W^*(s) = p_0 + p_0 \frac{\lambda[1 - B^*(s)]}{s - \lambda[1 - B^*(s)]}$$

$$= p_0 \frac{1}{1 - \frac{\lambda}{s}[1 - B^*(s)]}$$

As $p_0 = 1 - \rho$, we finally obtain

$$W^*(s) = \frac{1 - \rho}{1 - \rho\left[\dfrac{1 - B^*(s)}{s \bar{x}}\right]}$$

which is Eq. (5.106).

5.24.

PROBLEM 5.25.

Let us relate $\overline{s^k}$, the kth moment of the time in system to $\overline{N^k}$, the kth moment of the number in system.

(a) Show that Eq. (5.98) leads directly to Little's result, namely
$$\overline{N} = \lambda \overline{s} \triangleq \lambda T$$

(b) From Eq. (5.98) establish the second-moment relationship
$$\overline{N^2} - \overline{N} = \lambda^2 \overline{s^2}$$

(c) Prove that the general relationship is
$$\overline{N(N-1) \cdots (N-k+1)} = \lambda^k \overline{s^k}$$

SOLUTION

Eq. (5.98) is
$$Q(z) = S^*(\lambda - \lambda z)$$

(a) We know that $\overline{N} = \overline{q} = Q^{(1)}(1)$. Since
$$Q^{(1)}(z) = S^{*(1)}(\lambda - \lambda z)(-\lambda)$$
then
$$\overline{N} = S^{*(1)}(0)(-\lambda) = (-\overline{s})(-\lambda) = \lambda \overline{s} = \lambda T$$

(b) In a similar fashion we have
$$Q^{(2)}(z) = S^{*(2)}(\lambda - \lambda z)(-\lambda)^2$$
Thus
$$\overline{N^2} - \overline{N} = \overline{q^2} - \overline{q} = Q^{(2)}(1) = S^{*(2)}(0)(-\lambda)^2$$
$$\overline{N^2} - \overline{N} = (-1)^2 \overline{s^2} \lambda^2 = \lambda^2 \overline{s^2}$$

(c) In general
$$Q^{(k)}(1) = S^{*(k)}(0)(-\lambda)^k = (-1)^k \overline{s^k}(-\lambda)^k = \lambda^k \overline{s^k}$$

But $Q(z) = \sum_{j=0}^{\infty} p_j z^j$ and so
$$Q^{(k)}(z) = \sum_{j=k}^{\infty} j(j-1) \cdots (j-k+1) p_j z^{j-k}$$

5.25.

Thus
$$Q^{(k)}(1) = \sum_{j=k}^{\infty} j(j-1) \cdots (j-k+1) p_j$$
$$= \overline{N(N-1) \cdots (N-k+1)}$$

So
$$\overline{N(N-1) \cdots (N-k+1)} = \lambda^k \overline{s^k}.$$

5.25.

Chapter 6
The Queue G/M/m

PROBLEM 6.1.

Prove Eq. (6.13). [HINT: condition on an interarrival time of duration t and then further condition on the time ($\leq t$) it will take to empty the queue.]

SOLUTION

We wish to find p_{ij}, where $j < m < i+1$, by conditioning on the interarrival time t_{n+1} and also on $\tilde{y} \triangleq$ time for the queue to empty. We note that the queue becomes empty when $i+1-m$ have been served. Thus $\tilde{y}$ = time to serve $i+1-m$, which is the sum of $i+1-m$ intervals each exponentially distributed with parameter $m\mu$ (recall that the residual life of an exponential is distributed the same as the original interval). Thus $\tilde{y}$ has an $(i+1-m)$-stage Erlangian distribution with parameter $m\mu$. Therefore Eq. (2.147) gives the density of $\tilde{y}$ as

$$f_{\tilde{y}}(y) = \frac{m\mu(m\mu y)^{i-m}}{(i-m)!} e^{-m\mu y}$$

For the remaining interval, we note that $m-j$ of the m customers that are left must be served in $t_{n+1}-\tilde{y}$ seconds. Given that $t_{n+1} = t$ and $\tilde{y} = y$, this last occurs with probability

$$\binom{m}{j}\left[1 - e^{-\mu(t-y)}\right]^{m-j}\left[e^{-\mu(t-y)}\right]^{j}$$

Removing the condition on $\tilde{y}$ gives

$$p_{ij}\Big|_{t_{n+1}=t} = \int_0^t \binom{m}{j} e^{-\mu(t-y)j}\left[1 - e^{-\mu(t-y)}\right]^{m-j} f_{\tilde{y}}(y)\, dy$$

$$p_{ij}\Big|_{t_{n+1}=t} = \binom{m}{j} e^{-\mu t j} \int_0^t e^{\mu y j}\left[1 - e^{-\mu(t-y)}\right]^{m-j} \frac{m\mu(m\mu y)^{i-m}}{(i-m)!} e^{-m\mu y}\, dy$$

$$p_{ij}\Big|_{t_{n+1}=t} = \binom{m}{j} e^{-\mu t j} \int_0^t \frac{m\mu(m\mu y)^{i-m}}{(i-m)!} e^{-\mu y(m-j)}\left[1 - e^{-\mu(t-y)}\right]^{m-j} dy$$

Finally, removing the condition on t_{n+1} gives

$$p_{ij} = \int_0^\infty \binom{m}{j} e^{-j\mu t} \left[\int_0^t \frac{(m\mu y)^{i-m}}{(i-m)!} (e^{-\mu y} - e^{-\mu t})^{m-j} m\mu \, dy \right] dA(t)$$

PROBLEM 6.2.

Consider $E_2/M/1$ (with infinite queueing room).
(a) Solve for r_k in terms of σ.
(b) Evaluate σ explicitly.

SOLUTION

(a) By Eq. (6.27) for G/M/1

$$r_k = (1-\sigma)\sigma^k \qquad k = 0, 1, 2, \ldots$$

(b) Eq. (6.28) gives $\sigma = A^*(\mu - \mu\sigma)$ where $0 \leq \sigma < 1$. For $E_2/M/1$ we have $A^*(s) = \left(\dfrac{2\lambda}{s+2\lambda}\right)^2$. So $\sigma = \left(\dfrac{2\lambda}{\mu - \mu\sigma + 2\lambda}\right)^2$ or

$$\sigma[\mu^2(1 - 2\sigma + \sigma^2) + 4\lambda\mu(1-\sigma) + 4\lambda^2] = 4\lambda^2$$

$$\mu^2\sigma^3 - (2\mu^2 + 4\lambda\mu)\sigma^2 + (\mu^2 + 4\lambda\mu + 4\lambda^2)\sigma - 4\lambda^2 = 0$$

Since $\sigma = 1$ is always a root, we have

$$(\sigma - 1)[\mu^2\sigma^2 - \mu(\mu + 4\lambda)\sigma + 4\lambda^2] = 0$$

We now seek that root of the quadratic

$$\sigma^2 - \left(1 + \frac{4\lambda}{\mu}\right)\sigma + 4\left(\frac{\lambda}{\mu}\right)^2 = 0$$

for which $0 < \sigma < 1$. Since $\rho = \dfrac{\lambda}{\mu}$, $\sigma^2 - (1 + 4\rho)\sigma + 4\rho^2 = 0$. Thus

$$\sigma = \frac{1 + 4\rho \pm \sqrt{1 + 8\rho}}{2}$$

Since $\rho \geq 0$, $\dfrac{1 + 4\rho + \sqrt{1 + 8\rho}}{2} \geq 1$. Therefore the root we seek is

$$\sigma = \frac{1 + 4\rho - \sqrt{1 + 8\rho}}{2} \qquad \blacksquare$$

Let us show that $0 < \sigma < 1$: For $\rho > 0$, we clearly have
$$\sigma > \frac{1+4\rho-\sqrt{1+8\rho+16\rho^2}}{2}$$
or $\sigma > 0$. Also, for $\rho < 1$ we have $16\rho^2 < 16\rho$. Hence
$$16\rho^2 - 8\rho + 1 < 16\rho - 8\rho + 1 = 8\rho + 1$$
and so
$$4\rho - 1 \leqslant |4\rho - 1| < \sqrt{1+8\rho}$$
Thus $1+4\rho-\sqrt{1+8\rho} < 2$ which yields $\sigma < 1$. For a stable system we know that $0 < \rho < 1$ which, from the above argument, implies $0 < \sigma < 1$.

PROBLEM 6.3.

Consider M/M/m.
(a) How do p_k and r_k compare?
(b) Compare Eqs. (6.22) and (3.40).

SOLUTION

(a) For Poisson arrivals, we know $p_k = r_k$ (see Eq. (4.6)).
(b) Eq. (6.22) gives
$$P[\text{arrival queues}] = \sum_{k=m}^{\infty} r_k = K\frac{\sigma^m}{1-\sigma} \quad (G/M/m)$$
Eq. (3.40) gives
$$P[\text{queueing}] = \sum_{k=m}^{\infty} p_k = p_0\left(\frac{(m\rho)^m}{m!}\right)\left(\frac{1}{1-\rho}\right)$$
Since $p_k = r_k$ for all k, these two probabilities are the same.
[If we wish to solve for the constant K, we proceed as follows:
$$\sigma = A^*(m\mu - m\mu\sigma) = \frac{\lambda}{m\mu(1-\sigma)+\lambda}$$
and so $\sigma = \frac{\lambda}{m\mu} = \rho$. Thus equating the above probabilities we obtain $K = p_0\frac{m^m}{m!}$ where p_0 is given in Eq. (3.39).]

PROBLEM 6.4.

Prove Eq. (6.31).

SOLUTION

We have $W = \int_0^\infty y\, dW(y)$. Using Eq. (6.30) to determine $dW(y)$,

$$W = \int_0^\infty y\sigma\mu(1-\sigma) e^{-\mu(1-\sigma)y} dy = \sigma \frac{1}{\mu(1-\sigma)}$$

which is Eq. (6.31).

PROBLEM 6.5.

Show that Eq. (6.52) follows from Eq. (6.50).

SOLUTION

From Eq. (6.50), the left-hand side of Eq. (6.52) is

$$\sum_{i=2}^\infty \sigma^{i-1} p_{i1} = \sum_{i=2}^\infty \sigma^{i-1} \int_0^\infty \binom{2}{1} e^{-\mu t} \left[\int_0^t \frac{(2\mu y)^{i-2}}{(i-2)!} (e^{-\mu y} - e^{-\mu t}) 2\mu\, dy \right] dA(t)$$

Moving the summation inside both integrals, we have

$$\sum_{i=2}^\infty \sigma^{i-1} p_{i1} = 2\sigma \int_0^\infty e^{-\mu t} \left[\int_0^t \sum_{i=2}^\infty \frac{(2\mu y \sigma)^{i-2}}{(i-2)!} (e^{-\mu y} - e^{-\mu t}) 2\mu\, dy \right] dA(t)$$

Summing on i, we get

$$\sum_{i=2}^\infty \sigma^{i-1} p_{i1} = 2\sigma \int_0^\infty e^{-\mu t} \left[\int_0^t e^{2\mu y\sigma} (e^{-\mu y} - e^{-\mu t}) 2\mu\, dy \right] dA(t)$$

Carrying out the inner integration,

$$\sum_{i=2}^\infty \sigma^{i-1} p_{i1} = \frac{4\sigma}{2\sigma - 1} \int_0^\infty (e^{-2\mu(1-\sigma)t} - e^{-\mu t})\, dA(t)$$

$$- 2 \int_0^\infty (e^{-2\mu(1-\sigma)t} - e^{-2\mu t})\, dA(t)$$

Recognizing $A^*(\cdot)$, we may write

$$\sum_{i=2}^{\infty} \sigma^{i-1} p_{i1} = \frac{4\sigma}{2\sigma - 1} \left[A^*(2\mu(1-\sigma)) - A^*(\mu) \right] - 2 \left[A^*(2\mu(1-\sigma)) - A^*(2\mu) \right]$$

$$= 2A^*(2\mu) + \frac{2A^*(2\mu - 2\mu\sigma) - 4\sigma A^*(\mu)}{2\sigma - 1}$$

which is the right-hand side of Eq. (6.52).

PROBLEM 6.6.

Consider an $H_2/M/1$ system in which $\lambda_1 = 2$, $\lambda_2 = 1$, $\mu = 2$, and $\alpha_1 = 5/8$.
(a) Find σ.
(b) Find r_k.
(c) Find $w(y)$.
(d) Find W.

SOLUTION

(a) For $H_2/M/1$, $A^*(s) = \alpha_1 \dfrac{\lambda_1}{s + \lambda_1} + \alpha_2 \dfrac{\lambda_2}{s + \lambda_2}$ where $\alpha_2 = 1 - \alpha_1$. Thus

$$A^*(s) = \frac{5}{8} \cdot \frac{2}{s+2} + \frac{3}{8} \cdot \frac{1}{s+1}$$

$\sigma = A^*(\mu - \mu\sigma)$ for G/M/1, so using $\mu = 2$ we find

$$\sigma = \frac{5}{8} \cdot \frac{2}{2 - 2\sigma + 2} + \frac{3}{8} \cdot \frac{1}{2 - 2\sigma + 1}$$

or

$$\sigma(4 - 2\sigma)(3 - 2\sigma) = \frac{21}{4} - \frac{13}{4}\sigma$$

Thus

$$4\sigma^3 - 14\sigma^2 + \frac{61}{4}\sigma - \frac{21}{4} = 0$$

$\sigma = 1$ is always a root, so we factor

$$(\sigma - 1)\left[4\sigma^2 - 10\sigma + \frac{21}{4} \right] = 0$$

or
$$(\sigma-1)(4\sigma-7)\left(\sigma-\frac{3}{4}\right) = 0$$

The condition $0 < \sigma < 1$ gives
$$\sigma = \frac{3}{4}.$$ ∎

(b) For G/M/1, $r_k = (1-\sigma)\sigma^k$
$$\therefore r_k = \frac{1}{4}\left(\frac{3}{4}\right)^k$$ ∎

(c) Differentiating Eq. (6.30) gives
$$w(y) = (1-\sigma)u_0(y) + \sigma\mu(1-\sigma)e^{-\mu(1-\sigma)y} \quad y \geq 0$$

Thus
$$w(y) = \frac{1}{4}u_0(y) + \frac{3}{8}e^{-\frac{y}{2}} \quad y \geq 0$$

(d) By Eq. (6.31) $W = \dfrac{\sigma}{\mu(1-\sigma)}$. Thus
$$W = \frac{3/4}{2(1/4)} = \frac{3}{2}$$ ∎

PROBLEM 6.7.

Consider a D/M/1 system with $\mu = 2$ and with the same ρ as in the previous exercise.
(a) Find σ (correct to two decimal places).
(b) Find r_k.
(c) Find $w(y)$.
(d) Find W.

SOLUTION

(a) Since ρ and $\bar{x}$ ($= 1/\mu$) are the same as in Exercise (6.6), so is $\bar{t}$. Thus
$$\bar{t} = \frac{\alpha_1}{\lambda_1} + \frac{\alpha_2}{\lambda_2} = \frac{5}{16} + \frac{6}{16} = \frac{11}{16}$$

6.6. – 6.7.

For D/M/1, $A^*(s) = e^{-\bar{s}t}$ and so $A^*(s) = e^{-\frac{11}{16}s}$. Now

$$\sigma = A^*(\mu - \mu\sigma) = A^*(2 - 2\sigma) = e^{-\frac{11}{8}(1-\sigma)}$$

or

$$\log_e \sigma = \frac{11}{8}(\sigma - 1)$$

Since $\sigma - 1$ is tangent to $\log_e \sigma$ at $\sigma = 1$, we see that this last equation has exactly two roots.

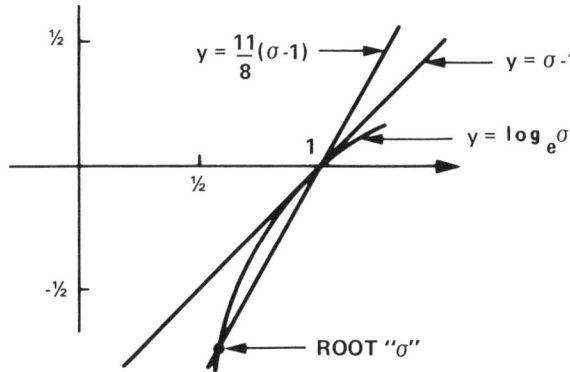

Disregarding the root at $\sigma = 1$, we find a root satisfying $0 < \sigma < 1$ numerically to obtain

$$\sigma \cong .51 \qquad \blacksquare$$

(b) For G/M/1, $r_k = (1 - \sigma)\sigma^k$. So

$$r_k \cong (.49)(.51)^k \qquad \blacksquare$$

(c) From Eq. (6.30) we have

$$w(y) = (1 - \sigma)u_0(y) + \sigma\mu(1 - \sigma)e^{-\mu(1-\sigma)y} \quad y \geq 0$$

Since $\mu = 2$ and $\sigma \cong .51$ we have

$$w(y) \cong (.49)u_0(y) + (.50)e^{-(.98)y} \quad y \geq 0 \qquad \blacksquare$$

(d) From Eq. (6.31) we have $W = \dfrac{\sigma}{\mu(1-\sigma)}$. Thus

$$W \cong .52 \qquad \blacksquare$$

6.7.

PROBLEM 6.8.

Consider a G/M/1 queueing system with room for at most two customers (one in service plus one waiting). Find r_k ($k=0,1,2$) in terms of μ and $A^*(s)$.

SOLUTION

Our task is to solve $\mathbf{r} = \mathbf{rP}$ where $\mathbf{r} = [r_0, r_1, r_2]$. First we must find the 3×3 matrix

$$\mathbf{P} = \begin{bmatrix} p_{00} & p_{01} & p_{02} \\ p_{10} & p_{11} & p_{12} \\ p_{20} & p_{21} & p_{22} \end{bmatrix}$$

Clearly $p_{02} = 0$. Furthermore, we now show that the last two rows of $\mathbf{P}$ are equal. If C_n arrives to find a full system ($q_n' = 2$) then he will be lost, whereas if he arrives to find one in the system ($q_n' = 1$) then he queues up and makes a full system. Due to the memoryless property of the service time distribution, these two cases are equivalent in terms of what the next customer, C_{n+1}, sees. Thus

$$p_{2i} = p_{1i} \quad \text{for} \quad i = 0, 1, 2$$

and so

$$\mathbf{P} = \begin{bmatrix} p_{00} & p_{01} & 0 \\ p_{10} & p_{11} & p_{12} \\ p_{10} & p_{11} & p_{12} \end{bmatrix}$$

Let us now find these matrix entries:

$$p_{01} = p_{12} = P[\text{service time} > \text{interarrival time}] = \int_0^\infty e^{-\mu t} dA(t) = A^*(\mu)$$

Thus also $p_{00} = 1 - p_{01} = 1 - A^*(\mu)$. To find p_{10} assume that $q_n' = 1$ (thus there will be 2 in the system) and that $q'_{n+1} = 0$. Thus we have

$$p_{10} = P[\text{residual life of } C_{n-1} \text{ plus service time of } C_n \leq \text{interarrival time}]$$

Since the service time is exponential, this sum is Erlangian; it has density $\mu(\mu x)e^{-\mu x}$ and distribution $1 - e^{-\mu x} - \mu x e^{-\mu x}$. Thus

$$p_{10} = \int_0^\infty (1 - e^{-\mu t} - \mu t e^{-\mu t}) \, dA(t)$$

6.8.

or

$$p_{10} = 1 - A^*(\mu) + \mu A^{*(1)}(\mu)$$

Finally

$$p_{11} = 1 - p_{10} - p_{12} = -\mu A^{*(1)}(\mu)$$

Thus

$$\mathbf{P} = \begin{bmatrix} 1 - A^*(\mu) & A^*(\mu) & 0 \\ 1 - A^*(\mu) + \mu A^{*(1)}(\mu) & -\mu A^{*(1)}(\mu) & A^*(\mu) \\ 1 - A^*(\mu) + \mu A^{*(1)}(\mu) & -\mu A^{*(1)}(\mu) & A^*(\mu) \end{bmatrix}$$

Now we determine r_k. From $\mathbf{r} = \mathbf{rP}$ ($[r_0, r_1, r_2] = [r_0, r_1, r_2]\mathbf{P}$) we have the two independent balance equations

$$r_1 = r_0 A^*(\mu) - (r_1 + r_2)\mu A^{*(1)}(\mu)$$

$$r_2 = (r_1 + r_2) A^*(\mu)$$

or, since $r_1 + r_2 = 1 - r_0$,

$$r_1 = r_0 A^*(\mu) - (1 - r_0)\mu A^{*(1)}(\mu)$$

$$r_2 = (1 - r_0) A^*(\mu)$$

The conservation of probability equation $r_0 + r_1 + r_2 = 1$ now yields

$$r_0 + \left[r_0 A^*(\mu) - (1 - r_0)\mu A^{*(1)}(\mu) \right] + \left[(1 - r_0) A^*(\mu) \right] = 1$$

Thus

$$r_0 = \frac{1 - A^*(\mu) + \mu A^{*(1)}(\mu)}{1 + \mu A^{*(1)}(\mu)} \quad \blacksquare$$

Using this result we also find

$$r_1 = \frac{\left[1 - A^*(\mu) \right] A^*(\mu)}{1 + \mu A^{*(1)}(\mu)} \quad \blacksquare$$

$$r_2 = \frac{\left[A^*(\mu) \right]^2}{1 + \mu A^{*(1)}(\mu)} \quad \blacksquare$$

6.8.

PROBLEM 6.9.

Consider a G/M/1 system in which the cost of making a customer wait y sec is
$$c(y) = ae^{by}$$
(a) Find the average cost of queueing for a customer.
(b) Under what conditions will the average cost be finite?

SOLUTION

For G/M/1,
$$W(y) = 1 - \sigma e^{-\mu(1-\sigma)y} \quad y \geq 0$$

Thus
$$w(y) = (1-\sigma)u_0(y) + \sigma\mu(1-\sigma)e^{-\mu(1-\sigma)y} \quad y \geq 0$$

(a) Average cost of queueing is
$$\bar{c} \triangleq \int_0^\infty c(y)\,dW(y) = \int_0^\infty ae^{by}w(y)\,dy$$
$$= a(1-\sigma) + \int_0^\infty a\sigma\mu(1-\sigma)e^{-[\mu(1-\sigma)-b]y}\,dy$$

Assuming $\mu(1-\sigma) - b > 0$ so that the integral exists, we find
$$\bar{c} = a(1-\sigma) + \frac{a\sigma\mu(1-\sigma)}{\mu(1-\sigma) - b}$$

or
$$\bar{c} = a(1-\sigma)\frac{\mu - b}{\mu(1-\sigma) - b} \quad \blacksquare$$

(b) The average cost is finite if
$$b < \mu(1-\sigma) \quad \blacksquare$$

as indicated in part (a).

Chapter 7

The Method of Collective Marks

PROBLEM 7.1.

Consider the M/G/1 system shown in the figure below with average arrival rate λ and service-time distribution $= B(x)$. Customers are served first-come-first-served from queue A until they either leave or receive a sec of service, at which time they join an entrance box as shown in the figure.

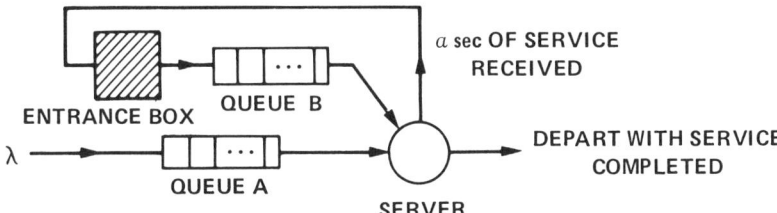

Customers continue to collect in the entrance box forming a group until queue A empties and the server becomes free. At this point, the entrance box "dumps" all it has collected as a *bulk arrival* to queue B. Queue B will receive service until a new arrival (to be referred to as a "starter") joins queue A at which time the server switches from queue B to serve queue A and the customer who is preempted returns to the head of queue B. The entrance box then begins to fill and the process repeats. Let

$g_n = P[\text{entrance box delivers bulk of size } n \text{ to queue } B]$

$$G(z) = \sum_{n=0}^{\infty} g_n z^n$$

(a) Give a probabilistic interpretation for $G(z)$ using the method of collective marks.

(b) Given that the "starter" reaches the entrance box, and using the method of collective marks find [in terms of λ, a, $B(\cdot)$, and $G(z)$]

$P_k = P[k$ customers arrive to queue A during the "starter's" service time and no marked customers arrive to the entrance box from the k sub-busy periods created in queue A by each of these customers]

7.1.

(c) Given that the "starter" does *not* reach the entrance box, find P_k as defined above.

(d) From (b) and (c), give an expression (involving an integral) for $G(z)$ in terms of λ, a, $B(\cdot)$, and itself.

(e) From (d) find the average bulk size $\bar{n} = \sum_{n=0}^{\infty} n g_n$.

SOLUTION

(a) Let $z = P[\text{customer is not marked upon arrival to the system}]$. Then

$$G(z) = \sum_{n=0}^{\infty} g_n z^n = P[\text{entrance box delivers no marked customers to } B]$$

$$= P[\text{no marked customer joins the entrance box during its filling time}]$$

$$= P[\text{no marked customer joins the entrance box during queue } A\text{'s busy period}]$$

(b) The probability that exactly k customers arrive to queue A during the "starter's" a seconds of service is simply $\dfrac{(\lambda a)^k}{k!} e^{-\lambda a}$. The probability that no marked customers arrive to the entrance box from the k sub-busy periods created in queue A is $[G(z)]^k$.

$$\therefore \quad P_k = \frac{[\lambda a G(z)]^k}{k!} e^{-\lambda a}$$

(c) Since the "starter" does not reach the entrance box, his service time $\tilde{x}$ is $\leqslant a$ (and has conditional density $\dfrac{dB(t)}{B(a)}$). Thus

$$P_k = \frac{1}{B(a)} \int_0^a \frac{[\lambda x G(z)]^k}{k!} e^{-\lambda x} dB(x)$$

(d) Let E be the event: the "starter" reaches the entrance box, and let E^C be the complementary event. Clearly $P[E] = 1 - B(a)$. Now

$G(z) = P[\text{no marked customer joins the entrance box}]$

$$= \left(G(z)\Big|_E\right) P[E] + \left(G(z)\Big|_{E^C}\right) P[E^C]$$

$$= \left(G(z)\Big|_E\right)[1 - B(a)] + \left(G(z)\Big|_{E^C}\right) B(a)$$

If E occurs, then the "starter" reaches the entrance box; he will not be marked with probability z and so

7.1.

$$G(z)\bigg|_E = \left(\sum_{k=0}^{\infty} \frac{[\lambda a G(z)]^k}{k!} e^{-\lambda a}\right) \cdot z$$

$$= z e^{-\lambda a [1 - G(z)]}$$

and

$$G(z)\bigg|_{E^C} = \frac{1}{B(a)} \sum_{k=0}^{\infty} \int_0^a \frac{[\lambda x G(z)]^k}{k!} e^{-\lambda x} dB(x)$$

$$= \frac{\int_0^a e^{-\lambda x [1 - G(z)]} dB(x)}{B(a)}$$

Thus

$$G(z) = z e^{-\lambda a [1 - G(z)]} [1 - B(a)] + \int_0^a e^{-\lambda x [1 - G(z)]} dB(x) \quad \blacksquare$$

(e) $\bar{n} = \sum_{n=0}^{\infty} n g_n = G^{(1)}(1)$. From (d) we have

$$G^{(1)}(z) = [1 - B(a)]\left[e^{-\lambda a [1 - G(z)]} + z e^{-\lambda a [1 - G(z)]} \lambda a G^{(1)}(z)\right]$$

$$+ \int_0^a e^{-\lambda x [1 - G(z)]} \lambda x G^{(1)}(z) \, dB(x)$$

Setting $z = 1$ gives

$$\bar{n} = G^{(1)}(1) = [1 - B(a)][1 + \lambda a G^{(1)}(1)] + \int_0^a \lambda x G^{(1)}(1) \, dB(x)$$

$$\bar{n} = [1 - B(a)](1 + \lambda a \bar{n}) + \int_0^a \lambda x \bar{n} \, dB(x)$$

$$\bar{n} = \frac{1 - B(a)}{1 - \lambda \left[\int_0^a x \, dB(x) + a[1 - B(a)]\right]} \quad \blacksquare$$

We note that the bracketed term in the denominator is simply the mean, $\bar{x}_a$, of the service time truncated at $\tilde{x} = a$.

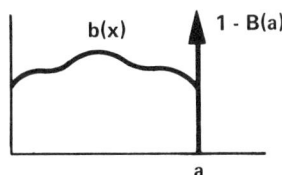

7.1.

Thus

$$\bar{n} = \frac{1-B(a)}{1-\lambda\bar{x}_a}$$ ∎

$$\bar{n} = \frac{1-B(a)}{1-\rho_a}$$ ∎

PROBLEM 7.2.

Consider the M/G/∞ system. We wish to find $P(z, t)$ as defined in Eq. (7.6). Assume the system contains $i=0$ customers at $t=0$. Let $p(t)$ be the probability that a customer who arrived in the interval $(0, t)$ is still present at t. Proceed as in Example 2 of Section 7.1.
(a) Express $p(t)$ in terms of $B(x)$.
(b) Find $P(z, t)$ in terms of λ, t, z, and $p(t)$.
(c) From (b) find $P_k(t)$ defined in Eq. (7.5).
(d) From (c), find $\lim P_k(t) = p_k$ as $t \to \infty$.

SOLUTION

(a) Since we have Poisson arrivals, a customer who arrived in $(0, t)$ has an arrival time τ which is uniformly distributed over $(0, t)$,

i.e. $P[\tau \leq x] = \dfrac{x}{t}$ for $0 \leq x \leq t$

Moreover,

$P[\text{customer still in service at } t \mid \text{arrived at } \tau] = 1 - B(t-\tau)$

Thus

$$p(t) = \int_0^t [1 - B(t-\tau)] \frac{d\tau}{t}$$

$$p(t) = \frac{1}{t}\int_0^t [1 - B(x)]\, dx$$ ∎

(b) Recall from page 263 (lines 2 and 3) that

$P(z, t)$ = probability that the system contains no marked customers at time t

7.1.–7.2.

Therefore

$$P(z,t) = \sum_{k=0}^{\infty} P\begin{bmatrix}\text{system contains no marked customers} \\ \text{at time } t \mid k \text{ arrivals in } (0,t)\end{bmatrix} \frac{(\lambda t)^k}{k!} e^{-\lambda t}$$

$$P(z,t) = \sum_{k=0}^{\infty} \left(1 - P\begin{bmatrix}\text{customer is present at time } t \\ \text{and is marked}\end{bmatrix}\right)^k \frac{(\lambda t)^k}{k!} e^{-\lambda t}$$

$$P(z,t) = \sum_{k=0}^{\infty} [1 - (1-z)p(t)]^k \frac{(\lambda t)^k}{k!} e^{-\lambda t}$$

$$P(z,t) = e^{-\lambda t(1-z)p(t)} \qquad \blacksquare$$

(c) From part (b) and Eq. (2.134), we see that $P(z,t)$ is the z-transform of a Poisson distribution with parameter $\lambda t\, p(t)$ and thus

$$P_k(t) = \frac{[\lambda t\, p(t)]^k}{k!} e^{-\lambda t\, p(t)} \qquad \blacksquare$$

(d)

$$p_k \triangleq \lim_{t\to\infty} P_k(t) = \lim_{t\to\infty} \frac{[\lambda t\, p(t)]^k}{k!} e^{-\lambda t\, p(t)}$$

But, using part (a),

$$\lim_{t\to\infty} t\, p(t) = \lim_{t\to\infty} \int_0^t [1-B(x)]\, dx = \int_0^\infty [1-B(x)]\, dx = \bar{x}$$

$$\therefore\ p_k = \frac{(\lambda \bar{x})^k}{k!} e^{-\lambda \bar{x}} \qquad \blacksquare$$

Note that this is Poisson with parameter $\lambda \bar{x}$. Observe how much simpler this approach is as compared to the derivation in Exercise (5.8).

PROBLEM 7.3.

Consider an M/G/1 queue, which is idle at time 0. Let $p = P[\text{no catastrophe occurs during the time the server is busy with those customers who arrived during } (0, t)]$ and let $q = P[\text{no catastrophe occurs during } (0, t + U(t))]$ where $U(t)$ is the unfinished work at time t. Catastrophes occur at a rate γ.
(a) Find p.
(b) Find q.
(c) Interpret $p - q$ as a probability and find an independent expression for it. We may then use (a) and (b) to relate the distribution of unfinished work to $B^*(s)$.

SOLUTION

We first define subsets B and I of the interval $(0, t+U(t))$ and events E_B, $E_{B \cup I}$ related to them.

Let B = the time the server is busy with those customers who arrived during $(0, t)$. So B = the time in $(0, t+U(t))$ when the server is busy. Let I = the time in $(0, t+U(t))$ when the server is idle. So $B \cup I = (0, t+U(t))$. Also let E_B = the event that no catastrophe occurs during B and $E_{B \cup I}$ = the event that no catastrophe occurs during $(0, t+U(t))$.

Clearly $E_{B \cup I} \subseteq E_B$. Also $p = P[E_B]$ and $q = P[E_{B \cup I}]$.

(a) We see, from Eq. (7.19), that $p = F^*(\gamma)$ where F^* is the Laplace transform for the distribution of $\tilde{t}_B$, the length (of time) of subset B. Since $\tilde{t}_B$ is a sum of service times of a random number of arrivals (the number of arrivals in $(0, t)$ from a Poisson process with parameter λ), we see from Eq. (II.34) that $F^*(s) = N(B^*(s))$ where $N(z) = e^{-\lambda t(1-z)}$ by Eq. (2.134). Thus

$$p = F^*(\gamma) = N(B^*(\gamma))$$

or

$$p = e^{-\lambda t [1 - B^*(\gamma)]} \qquad \blacksquare$$

(b) If we assume that $U(t) = u$, then

$$q|_{U(t)=u} = P[\text{no catastrophe in } (0, t+u)] = e^{-\gamma(t+u)}$$

Unconditioning, we find

$$q = \int_0^\infty e^{-\gamma(t+u)} \, d_u P[U(t) \leq u]$$

$$q = e^{-\gamma t} \int_0^\infty e^{-\gamma u} \, d_u P[U(t) \leq u] \qquad \blacksquare$$

(c) Recall $E_{B \cup I} \subseteq E_B$. The set diagram is

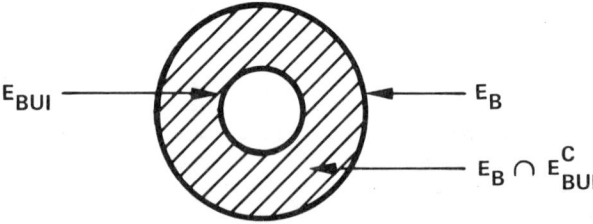

7.3.

From the diagram we see that $p - q = P[E_B \cap E_{B \cup I}^C]$. So $p - q = P[$no catastrophe occurs in B but at least one catastrophe occurs in $B \cup I]$. Thus $p - q = P[$at least one catastrophe occurs in the idle portion I of $(0, t + U(t))$ and no catastrophe occurs in the busy portion B of this interval$]$.

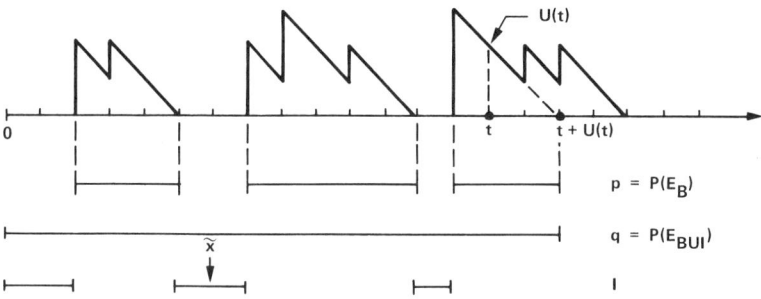

We now condition on the time $\tilde{x}$ of the first catastrophe (which necessarily occurs during an idle period).

$(p - q)\big|_{\tilde{x}=x} = P[$first catastrophe occurs at $\tilde{x} = x$ and no catastrophe occurs during the time the server is busy with arrivals in $(x, t)]$

i.e. since no catastrophe occurs prior to x, we need only concern ourselves with the busy intervals from x to $t + U(t)$. Unconditioning gives

$$p - q = \left[\int_0^t \gamma e^{-\gamma x} dx \right] P[U(x) = 0]\, e^{-\lambda(t-x)[1 - B^*(\gamma)]} \qquad \blacksquare$$

where we use the result from part (a) to obtain the final term. Hence, equating expressions for $p - q$ from (a), (b) and (c) gives

$$e^{-\lambda t[1 - B^*(\gamma)]} - e^{-\gamma t} \int_0^\infty e^{-\gamma u}\, d_u P[U(t) \leq u] =$$
$$\int_0^t \gamma e^{-\gamma x}\, dx\, P[U(x) = 0]\, e^{-\lambda(t-x)[1 - B^*(\gamma)]} \qquad \blacksquare$$

7.3.

PROBLEM 7.4.

Consider the G/M/m system. The root σ, which is defined in Eq. (6.21) plays a central role in the solution. Examine Eq. (6.21) from the viewpoint of collective marks and give a probabilistic interpretation for σ.

SOLUTION

We first give two interpretations of Eq. (6.21), one based on Section 7.1 and the second based on Section 7.2. Neither interpretation is very satisfying however. We then give a more interesting probabilistic interpretation for σ.
Note that

$$A^*(m\mu - m\mu\sigma) = \int_0^\infty e^{-m\mu(1-\sigma)t} dA(t)$$

$$= \int_0^\infty e^{-m\mu t} \sum_{k=0}^\infty \frac{(m\mu\sigma t)^k}{k!} dA(t)$$

$$= \sum_{k=0}^\infty \sigma^k \int_0^\infty e^{-m\mu t} \frac{(m\mu t)^k}{k!} dA(t)$$

The integral $\int_0^\infty e^{-m\mu t} \frac{(m\mu t)^k}{k!} dA(t)$ is simply the probability that exactly k customers complete service during an interarrival time in which all m servers remain busy. Now let customers be marked with probability $1 - \sigma$. Then, from the equation above,

$$A^*(m\mu - m\mu\sigma) = P[\text{only unmarked customers are served} \\ \text{in an interarrival time when} \\ \text{all } m \text{ servers remain busy}].$$

Eq. (6.21) then states that $\sigma = A^*(m\mu - m\mu\sigma)$; thus the probability of not marking a customer is defined in terms of itself through this last equation.
A second interpretation follows from Eq. (7.19). Thus $A^*(m\mu - m\mu\sigma)$ is the probability that an interarrival time (the event) occurs before the catastrophe, where catastrophes occur according to a Poisson process with rate $m\mu(1 - \sigma)$. Eq. (6.21) says that this probability is in fact σ, which again expresses σ in terms of itself.
Neither of these is very interesting. We point out (and invite the reader to supply a proof of this statement) that σ is also the conditional probability that a customer (say C_n) who must join the queue is still waiting in the queue when the next customer (C_{n+1}) arrives.

7.4.

Chapter 8

The Queue G/G/1

PROBLEM 8.1.

From Eq. (8.18) show that $C^*(s) = A^*(-s)B^*(s)$.

SOLUTION

Eq. (8.18) gives $c(u) = a(-u) \circledast b(u)$. From the definition of convolution (see also Eq. (8.17)) we have

$$c(u) = \int_0^\infty b(u+t)a(t)\,dt$$

We form the Laplace transform as

$$C^*(s) \triangleq \int_{-\infty}^\infty c(u)e^{-su}\,du$$

$$= \int_{-\infty}^\infty \left[\int_0^\infty b(u+t)a(t)\,dt\right]e^{-su}\,du$$

$$= \int_0^\infty a(t)e^{st}\left[\int_{-\infty}^\infty b(u+t)e^{-s(u+t)}\,du\right]dt$$

Let $x = u + t$, $dx = du$

$$C^*(s) = \int_0^\infty a(t)e^{st}\left[\int_{-\infty}^\infty b(x)e^{-sx}\,dx\right]dt$$

Since $b(x) = 0$ for $x < 0$,

$$C^*(s) = \int_0^\infty a(t)e^{st}\left[\int_0^\infty b(x)e^{-sx}\,dx\right]dt$$

$$C^*(s) = \int_0^\infty a(t)e^{st}\,dt \int_0^\infty b(x)e^{-sx}\,dx$$

$$C^*(s) = A^*(-s)B^*(s)$$

180

PROBLEM 8.2.

Find $C(u)$ for M/M/1.

SOLUTION

Recall that $C(u) = \int_0^\infty B(u+t)a(t)\,dt$. For M/M/1,
$$B(x) = \begin{cases} 1 - e^{-\mu x} & x \geq 0 \\ 0 & x < 0 \end{cases}$$

and
$$a(t) = \begin{cases} \lambda e^{-\lambda t} & t \geq 0 \\ 0 & t < 0 \end{cases}$$

To determine the limits of integration for $C(u)$, we consider two cases:
Case (1): $u \geq 0$

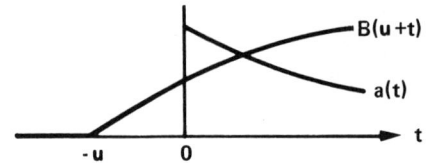

$$C(u) = \int_0^\infty (1 - e^{-\mu(u+t)}) \lambda e^{-\lambda t}\,dt$$
$$= \int_0^\infty \lambda e^{-\lambda t}\,dt - \int_0^\infty e^{-\mu u} \lambda e^{-(\lambda + \mu)t}\,dt$$
$$C(u) = 1 - \frac{\lambda}{\lambda + \mu} e^{-\mu u} \qquad \blacksquare$$

Case (2): $u < 0$

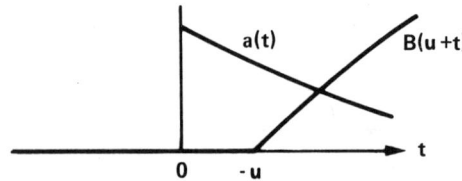

8.2.

$$C(u) = \int_{-u}^{\infty} (1 - e^{-\mu(u+t)}) \lambda e^{-\lambda t} \, dt$$

$$= \int_{-u}^{\infty} \lambda e^{-\lambda t} \, dt - \int_{-u}^{\infty} e^{-\mu u} \lambda e^{-(\lambda + \mu)t} \, dt$$

$$= e^{\lambda u} - e^{-\mu u} \left[\frac{\lambda}{\lambda + \mu} e^{(\lambda + \mu)u} \right]$$

$$C(u) = \frac{\mu}{\lambda + \mu} e^{\lambda u} \qquad \blacksquare$$

So

$$C(u) = \begin{cases} 1 - \dfrac{\lambda}{\lambda + \mu} e^{-\mu u} & u \geq 0 \\ \dfrac{\mu}{\lambda + \mu} e^{\lambda u} & u < 0 \end{cases}$$

We sketch $C(u)$ as follows:

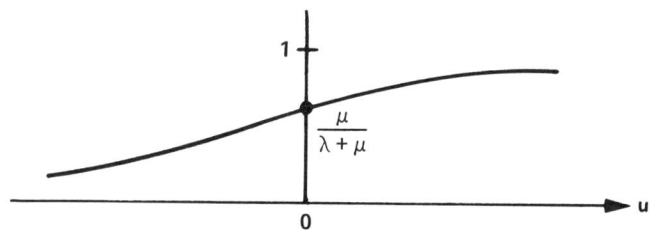

PROBLEM 8.3.

Consider the system M/D/1 with a fixed service time of $\bar{x}$ sec.
(a) Find

$$C(u) = P[u_n \leq u]$$

and sketch its shape.
(b) Find $E[u_n]$.

SOLUTION

(a) From the definition of u_n (Eq. (8.3)) we have
$$C(u) = P[x_n - t_{n+1} \leq u]$$
Since we have an M/D/1 system, then
$$C(u) = P[\bar{x} - u \leq t_{n+1}]$$
$$= 1 - P[t_{n+1} \leq \bar{x} - u]$$
and so $C(u) = 1 - A(\bar{x} - u)$. We also know that
$$A(t) = \begin{cases} 0 & t < 0 \\ 1 - e^{-\lambda t} & t \geq 0 \end{cases}$$
Thus
$$C(u) = \begin{cases} 1 & \bar{x} < u \\ e^{-\lambda(\bar{x}-u)} & \bar{x} \geq u \end{cases} \blacksquare$$

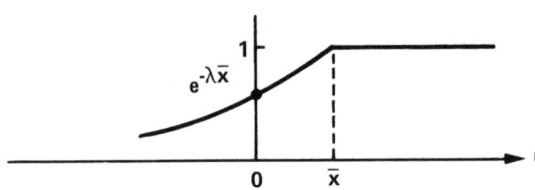

(b)
$$E[u_n] = E[x_n - t_{n+1}] = E[x_n] - E[t_{n+1}]$$
$$E[u_n] = \bar{x} - \bar{t} = \bar{t}(\rho - 1) \blacksquare$$

(true also for G/G/1)

PROBLEM 8.4.

For the sequence of random variables given below, generate the figure corresponding to Figure 8.3 and complete the table.

8.3. – 8.4.

n	0	1	2	3	4	5	6	7	8	9
t_{n+1}	2	1	1	5	7	2	2	1	1	6	$\cdots$
x_n	3	4	2	3	3	4	2	1	1	3	$\cdots$
u_n										
w_n measured										
w_n calculated										

SOLUTION

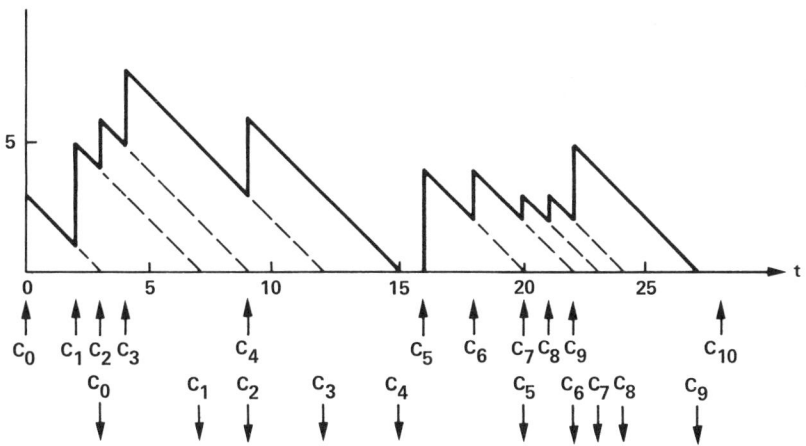

n	0	1	2	3	4	5	6	7	8	9
t_{n+1}	2	1	1	5	7	2	2	1	1	6	$\cdots$
x_n	3	4	2	3	3	4	2	1	1	3	$\cdots$
u_n	1	3	1	-2	-4	2	0	0	0	-3	$\cdots$
w_n measured	0	1	4	5	3	0	2	2	2	2	$\cdots$
w_n calculated	0	1	4	5	3	0	2	2	2	2	$\cdots$

w_n may also be measured from the graph as the value of $U(t)$ just before C_n enters the system.

8.4.

PROBLEM 8.5.†

Consider the case where $\rho = 1-\epsilon$ for $0 < \epsilon \ll 1$. Let us expand $W(y-u)$ in Eq. (8.23) as

$$W(y-u) = W(y) - uW^{(1)}(y) + \frac{u^2}{2}W^{(2)}(y) + R(u,y)$$

where $W^{(n)}(y)$ is the nth derivative of $W(y)$ and $R(u,y)$ is such that $\int_{-\infty}^{\infty} R(u,y)\,dC(u)$ is negligible due to the slow variation of $W(y)$ when $\rho = 1-\epsilon$. Let $\overline{u^k}$ denote the kth moment of $\tilde{u}$.

(a) Under these conditions convert Lindley's integral equation to a second-order linear differential equation involving $\overline{u^2}$ and $\bar{u}$.
(b) With the boundary condition $W(0) = 0$, solve the equation found in (a) and express the mean wait W in terms of the first two moments of $\tilde{t}$ and $\tilde{x}$.

SOLUTION

(a) Lindley's integral equation (Eq. (8.23)) is

$$W(y) = \begin{cases} \int_{-\infty}^{y} W(y-u)\,dC(u) & y \geq 0 \\ 0 & y < 0 \end{cases}$$

Integrating both sides of the expansion for $W(y-u)$ given in the exercise over all u, assuming $\int_{-\infty}^{\infty} R(u,y)\,dC(u)$ is negligible, and noting that

$$W(y) = \int_{-\infty}^{\infty} W(y-u)\,dC(u) \quad \text{for } y \geq 0$$

(since $W(y-u)$ is zero for $u > y$), we have

$$W(y) = \int_{-\infty}^{\infty} \left[W(y) - uW^{(1)}(y) + \frac{u^2}{2}W^{(2)}(y) \right] dC(u)$$

or

$$W(y) = W(y) - W^{(1)}(y)\bar{u} + W^{(2)}(y)\frac{\overline{u^2}}{2}$$

The required differential equation is therefore

$$\frac{\overline{u^2}}{2} W^{(2)}(y) - \bar{u} W^{(1)}(y) = 0 \qquad \blacksquare$$

8.5.

(b) Integrating the above equation yields the first-order equation

$$\frac{\overline{u^2}}{2} W^{(1)}(y) - \overline{u} W(y) = C$$

The homogeneous solution is

$$W_h(y) = K e^{(2\overline{u}/\overline{u^2})y}$$

Substituting back into our differential equation, we see that the particular solution must be

$$W_p(y) = -\frac{C}{\overline{u}}$$

Thus

$$W(y) = K e^{(2\overline{u}/\overline{u^2})y} - \frac{C}{\overline{u}}$$

The boundary condition $W(0) = 0$ gives $\dfrac{C}{\overline{u}} = K$. Thus

$$W(y) = K(e^{(2\overline{u}/\overline{u^2})y} - 1)$$

Now, as $y \to \infty$ we must have $W(y) \to 1$ (it is a PDF). Since $\dfrac{2\overline{u}}{\overline{u^2}} < 0$ for a stable system,

$$\lim_{y \to \infty} e^{(2\overline{u}/\overline{u^2})y} = 0$$

and so $K = -1$. Thus

$$W(y) = 1 - e^{(2\overline{u}/\overline{u^2})y} \quad y \geq 0 \qquad \blacksquare$$

The mean wait W is simply

$$W = -\frac{\overline{u^2}}{2\overline{u}} \qquad \blacksquare$$

From Eq. (8.13) and Eq. (8.95) we may rewrite W as

$$W = \frac{\sigma_a^2 + \sigma_b^2}{2\overline{t}(1-\rho)} + \frac{1}{2}\overline{t}(1-\rho) \quad (\rho \to 1) \qquad \blacksquare$$

8.5.

PROBLEM 8.6.

Consider the $D/E_r/1$ queueing system, with a constant interarrival time (of $\bar{t}$ sec) and a service-time pdf given as in Eq. (4.16).
(a) Find $C(u)$.
(b) Show that Lindley's integral equation yields $W(y - \bar{t}) = 0$ for $y < \bar{t}$ and

$$W(y - \bar{t}) = \int_0^y W(y - w) \, dB(w) \quad \text{for } y \geq \bar{t}$$

(c) Assume the following solution for $W(y)$:

$$W(y) = 1 + \sum_{i=1}^r a_i e^{\alpha_i y} \quad y \geq 0$$

where a_i and α_i may both be complex, but where $\text{Re}(\alpha_i) < 0$ for $i = 1, 2, \ldots, r$. Using this assumed solution, show that the following equations must hold:

$$e^{-\alpha_i \bar{t}} = \left(\frac{r\mu}{r\mu + \alpha_i}\right)^r \quad i = 1, 2, \ldots, r$$

$$\sum_{i=0}^r \frac{a_i}{(r\mu + \alpha_i)^{j+1}} = 0 \quad j = 0, 1, \ldots, r - 1$$

where $a_0 = 1$ and $\alpha_0 = 0$. Note that $\{\alpha_i\}$ may be found from the first set of (transcendental) equations, and then the second set gives $\{a_i\}$. It can be shown that the α_i are distinct. See [SYSK 60].

SOLUTION

(a) Eq. (8.17) gives the following formula for $C(u)$

$$C(u) = \int_0^\infty B(u + t) \, dA(t)$$

For $D/E_r/1$, $dA(t) = a(t) \, dt = u_0(t - \bar{t}) \, dt$. So $C(u) = B(u + \bar{t})$. Now Eq. (4.16) gives the service time density as

$$b(x) = \frac{r\mu(r\mu x)^{r-1}}{(r-1)!} e^{-r\mu x} \quad x \geq 0$$

The corresponding distribution function (obtained through repeated integration by parts) is

$$B(x) = 1 - \sum_{i=0}^{r-1} \frac{(r\mu x)^i}{i!} e^{-r\mu x} \quad x \geq 0$$

8.6.

Therefore

$$C(u) = B(u+\bar{t}) = \begin{cases} 0 & u < -\bar{t} \\ 1 - \sum_{i=0}^{r-1} \frac{[r\mu(u+\bar{t})]^i}{i!} e^{-r\mu(u+\bar{t})} & u \geq -\bar{t} \end{cases}$$

∎

(b) Using Eq. (8.23), evaluating it at $y - \bar{t}$, and then recalling that $C(u) = B(u+\bar{t})$, we find

$$W(y-\bar{t}) = \begin{cases} \int_{-\infty}^{y-\bar{t}} W(y-\bar{t}-u)\, dB(u+\bar{t}) & y \geq \bar{t} \\ 0 & y < \bar{t} \end{cases}$$

Substituting $w = u + \bar{t}$ gives

$$W(y-\bar{t}) = \begin{cases} \int_{-\infty}^{y} W(y-w)\, dB(w) & y \geq \bar{t} \\ 0 & y < \bar{t} \end{cases}$$

as desired.

(c) We first rewrite the assumed solution for $W(y)$ in the form

$$W(y) = \sum_{i=0}^{r} a_i e^{\alpha_i y} \quad y \geq 0$$

where we define $a_0 = 1$ and $\alpha_0 = 0$. For $y \geq \bar{t}$, we will substitute this assumed solution into the equation from part (b) to yield the desired relations. The left-hand side (LHS) of the equation becomes

$$\text{LHS} = W(y-\bar{t}) = \sum_{i=0}^{r} a_i e^{\alpha_i(y-\bar{t})}$$

The right-hand side (RHS) is

$$\text{RHS} = \int_0^y W(y-w)\, dB(w) = \int_0^y \sum_{i=0}^{r} a_i e^{\alpha_i(y-w)}\, dB(w)$$

$$= \sum_{i=0}^{r} a_i e^{\alpha_i y} \int_0^y e^{-\alpha_i w}\, dB(w)$$

$$\text{RHS} = \sum_{i=0}^{r} a_i e^{\alpha_i y} \int_0^y \frac{r\mu(r\mu w)^{r-1}}{(r-1)!} e^{-(r\mu+\alpha_i)w}\, dw$$

8.6.

Repeated integration by parts gives

$$\text{RHS} = \sum_{i=0}^{r} a_i e^{\alpha_i y} \left[-\sum_{j=1}^{r} \frac{(r\mu w)^{r-j}}{(r-j)!} \left(\frac{r\mu}{r\mu+\alpha_i} \right)^j e^{-(r\mu+\alpha_i)w} \right]_0^y$$

$$= \sum_{i=0}^{r} a_i e^{\alpha_i y} \left[\left(\frac{r\mu}{r\mu+\alpha_i} \right)^r - \sum_{j=1}^{r} \frac{(r\mu y)^{r-j}}{(r-j)!} \left(\frac{r\mu}{r\mu+\alpha_i} \right)^j e^{-(r\mu+\alpha_i)y} \right]$$

$$= \sum_{i=0}^{r} a_i e^{\alpha_i y} \left(\frac{r\mu}{r\mu+\alpha_i} \right)^r - \sum_{j=1}^{r} \frac{(r\mu)^r y^{r-j}}{(r-j)!} \sum_{i=0}^{r} \frac{a_i e^{-r\mu y}}{(r\mu+\alpha_i)^j}$$

Equating coefficients of $e^{\alpha_i y}$ on LHS and RHS gives

$$e^{-\alpha_i \bar{t}} = \left(\frac{r\mu}{r\mu+\alpha_i} \right)^r \quad 1 \leq i \leq r$$

while the coefficient of $e^{-r\mu y}$ must be zero, and so

$$\sum_{j=1}^{r} \frac{r\mu^r y^{r-j}}{(r-j)!} \sum_{i=0}^{r} \frac{a_i}{(r\mu+\alpha_i)^j} = 0 \quad \begin{pmatrix} \text{for all} \\ y \geq 0 \end{pmatrix}$$

Thus

$$\sum_{i=0}^{r} \frac{a_i}{(r\mu+\alpha_i)^j} = 0 \quad 1 \leq j \leq r$$

or

$$\sum_{i=0}^{r} \frac{a_i}{(r\mu+\alpha_i)^{j+1}} = 0 \quad 0 \leq j \leq r-1$$

PROBLEM 8.7.

Consider the following queueing systems in which *no queue* is permitted. Customers who arrive to find the system busy must leave without service.
(a) M/M/1: Solve for $p_k = P[k$ in system].
(b) M/H$_2$/1: As in Figure 4.10 with $\alpha_1 = \alpha$, $\alpha_2 = 1 - \alpha$, $\mu_1 = 2\mu\alpha$ and $\mu_2 = 2\mu(1-\alpha)$.
  (i) Find the mean service time $\bar{x}$.
  (ii) Solve for p_0 (an empty system), p_α (a customer in the $2\mu\alpha$ box) and $p_{1-\alpha}$ (a customer in the $2\mu(1-\alpha)$ box).
(c) H$_2$/M/1: Where $A(t)$ is hyperexponential as in (b), but with parameters $\mu_1 = 2\lambda\alpha$ and $\mu_2 = 2\lambda(1-\alpha)$ instead. Draw the state-transition diagram

(with labels on branches) for the following four states: E_{ij} is state with "arriving" customer in arrival stage i and j customers in service $i = 1, 2$ and $j = 0, 1$.

(d) M/E$_r$/1: Solve for $P_j = P[j$ stages of service left to go].

(e) M/D/1: With all service times equal to $\bar{x}$
   (i) Find the probability of an empty system.
   (ii) Find the fraction of lost customers.

(f) E$_2$/M/1: Define the four states as E_{ij} where i is the number of "arrival" stages left to go and j is the number of customers in service. Draw the labeled state-transition diagram.

SOLUTION

(a) M/M/1

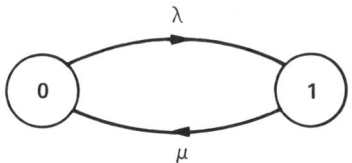

$$\lambda p_0 = \mu p_1 \text{ and } p_0 + p_1 = 1$$

$$\therefore p_1 = \frac{\lambda}{\lambda + \mu} \qquad \blacksquare$$

$$p_0 = \frac{\mu}{\lambda + \mu} \qquad \blacksquare$$

(b) M/H$_2$/1 (see Fig. 4.12)

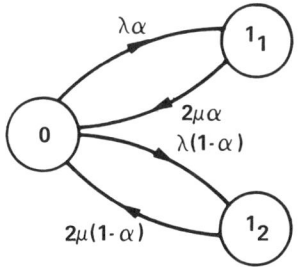

8.7.

(i)
$$\bar{x} = \alpha\left(\frac{1}{2\mu\alpha}\right) + (1-\alpha)\left(\frac{1}{2\mu(1-\alpha)}\right) = \frac{1}{\mu}$$ ∎

(ii)
$$\lambda\alpha p_0 = 2\mu\alpha p_\alpha$$

$$\lambda(1-\alpha)p_0 = 2\mu(1-\alpha)p_{1-\alpha}$$

$$p_0 + p_\alpha + p_{1-\alpha} = 1$$

$$\therefore p_0 = \frac{\mu}{\lambda+\mu}$$ ∎

$$p_\alpha = p_{1-\alpha} = \frac{\lambda}{2(\lambda+\mu)}$$ ∎

(c) $H_2/M/1$

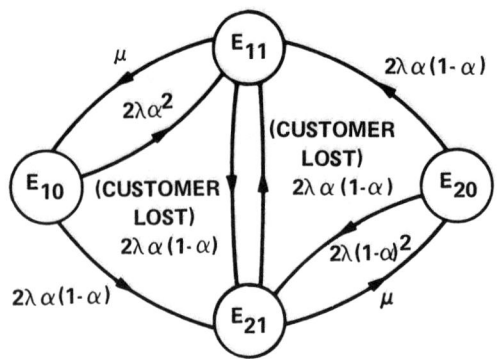

(d) $M/E_r/1$

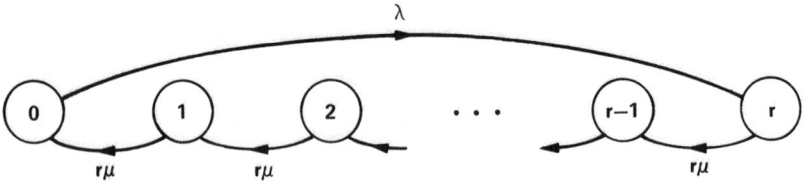

8.7.

The equilibrium equations for P_j are

$$\lambda P_0 = r\mu P_1$$

$$r\mu P_1 = r\mu P_2$$

$$\vdots$$

$$r\mu P_{r-1} = r\mu P_r$$

These equations plus the conservation of probability equation

$$P_0 + P_1 + \cdots + P_r = 1$$

yield

$$P_0 = \frac{\mu}{\lambda + \mu}$$

$$P_j = \frac{\lambda}{r(\lambda + \mu)} \quad 1 \leq j \leq r$$

(e) M/D/1
 (i) Using Eq. (4.18), M/D/1 = $\lim_{r \to \infty}$ M/E$_r$/1. So, by part (d),

$$P[\text{empty system}] = \frac{\mu}{\lambda + \mu} = \frac{1/\bar{x}}{\lambda + 1/\bar{x}} = \frac{1}{1 + \lambda \bar{x}}$$

[Note: The probability of an empty system for an M/G/1 system with no queueing may be found by the following renewal theoretic argument: Since we have Poisson arrivals, the time between a departure and a new arrival is exponentially distributed with mean $1/\lambda$. Thus the mean length of an idle period, say $\bar{I}$, is $1/\lambda$. Since no queue is allowed, the mean length of a busy period, say $\bar{B}$, is $\bar{x}$. By renewal theory,

$$P[\text{empty system}] = \frac{\bar{I}}{\bar{I} + \bar{B}} = \frac{1/\lambda}{1/\lambda + \bar{x}} = \frac{1}{1 + \lambda \bar{x}}$$

for M/G/1 with no queueing. Thus portions of parts (a), (b), (d) and (e) are special cases of this result.]

 (ii) The fraction of lost customers = probability of an arrival to a busy system = probability of a busy system (Poisson arrivals)

$$= 1 - p_0 = \frac{\lambda \bar{x}}{1 + \lambda \bar{x}}$$

8.7.

(f) $E_2/M/1$

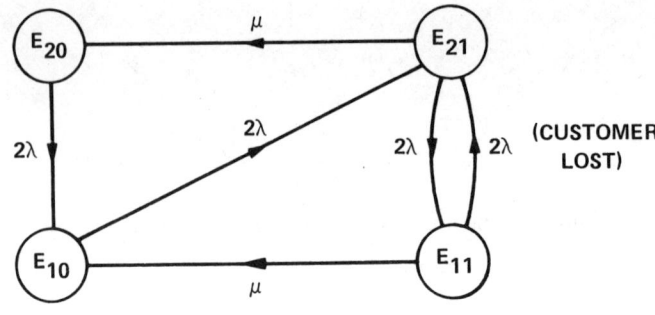

PROBLEM 8.8.

Consider a single-server queueing system in which the interarrival time is chosen with probability α from an exponential distribution of mean $1/\lambda$ and with probability $1-\alpha$ from an exponential distribution with mean $1/\mu$. Service is exponential with mean $1/\mu$.
- **(a)** Find $A^*(s)$ and $B^*(s)$.
- **(b)** Find the expression for $\Psi_+(s)/\Psi_-(s)$ and show the pole-zero plot in the s-plane.
- **(c)** Find $\Psi_+(s)$ and $\Psi_-(s)$.
- **(d)** Find $\Phi_+(s)$ and $W(y)$.

SOLUTION

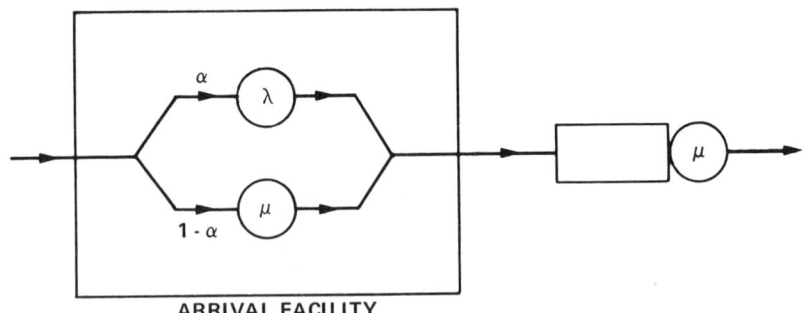

ARRIVAL FACILITY

8.7.–8.8.

We note that this is an $H_2/M/1$ system. Also, $\bar{x} = \dfrac{1}{\mu}$ and $\bar{t} = \dfrac{\alpha}{\lambda} + \dfrac{1-\alpha}{\mu}$ or $\bar{t} = \dfrac{1}{\mu} + \alpha\left(\dfrac{1}{\lambda} - \dfrac{1}{\mu}\right)$. Hence the system is stable if and only if

$$\alpha > 0 \quad \text{and} \quad 0 < \frac{\lambda}{\mu} < 1.$$

(a)

$$a(t) = \alpha\lambda e^{-\lambda t} + (1-\alpha)\mu e^{-\mu t} \quad t \geq 0$$

$$b(x) = \mu e^{-\mu x} \quad x \geq 0$$

Thus

$$A^*(s) = \frac{\alpha\lambda}{s+\lambda} + \frac{(1-\alpha)\mu}{s+\mu} \qquad \blacksquare$$

$$B^*(s) = \frac{\mu}{s+\mu} \qquad \blacksquare$$

(b)

$$\frac{\Psi_+(s)}{\Psi_-(s)} = A^*(-s)B^*(s) - 1 = \left[\frac{\alpha\lambda}{\lambda-s} + \frac{(1-\alpha)\mu}{\mu-s}\right]\frac{\mu}{s+\mu} - 1$$

$$\frac{\Psi_+(s)}{\Psi_-(s)} = \frac{(-\alpha\lambda s + \mu\lambda - \mu s + \alpha\mu s)\mu - (\lambda-s)(\mu^2 - s^2)}{(\lambda-s)(\mu-s)(\mu+s)}$$

$$\frac{\Psi_+(s)}{\Psi_-(s)} = \frac{-s[s^2 - \lambda s - \alpha\mu(\mu-\lambda)]}{(\lambda-s)(\mu-s)(\mu+s)}$$

The roots of $s^2 - \lambda s - \alpha\mu(\mu-\lambda)$ are

$$s_1 \triangleq \frac{\lambda + \sqrt{\lambda^2 + 4\alpha\mu(\mu-\lambda)}}{2}$$

$$s_2 \triangleq \frac{\lambda - \sqrt{\lambda^2 + 4\alpha\mu(\mu-\lambda)}}{2}$$

$$\frac{\Psi_+(s)}{\Psi_-(s)} = \frac{-s(s-s_1)(s-s_2)}{(\lambda-s)(\mu-s)(\mu+s)} \qquad \blacksquare$$

To locate the relative positions of the poles and zeroes, we note that for a stable system ($\alpha > 0$, $\lambda < \mu$), then $s_1 > \dfrac{\lambda + \sqrt{\lambda^2}}{2} = \lambda$. Further, since also $\alpha \leq 1$, then $s_1 \leq \dfrac{\lambda + \sqrt{\lambda^2 + 4\mu(\mu-\lambda)}}{2}$ or $s_1 \leq \dfrac{\lambda + (2\mu-\lambda)}{2} = \mu$. So we have $\lambda < s_1 \leq \mu$. Similarly, $\lambda - \mu \leq s_2 < 0$.

8.8.

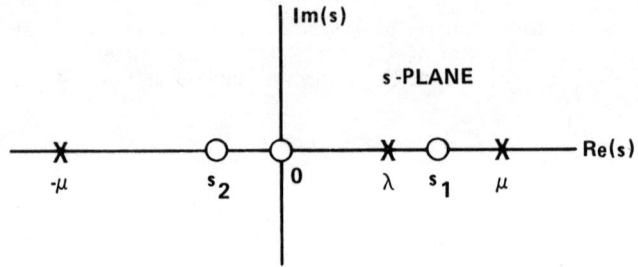

Zeroes (O) and poles (×) of $\Psi_+(s)/\Psi_-(s)$

(c) Using (8.36) and (8.37) we find

$$\Psi_+(s) = s\left(\frac{s-s_2}{s+\mu}\right)$$

$$\Psi_-(s) = -\frac{(\lambda-s)(\mu-s)}{s-s_1}$$

(d) Eq. (8.41) gives $\Phi_+(s) = \frac{K}{\Psi_+(s)}$. By Eq. (8.43) $K = \lim_{s\to 0}\frac{\Psi_+(s)}{s}$

$$K = \lim_{s\to 0}\frac{s-s_2}{s+\mu} = \frac{-s_2}{\mu}$$

$$\therefore \Phi_+(s) = -\frac{s_2}{\mu}\frac{s+\mu}{s(s-s_2)}$$

Next we invert this transform using partial fraction expansion techniques:

$$\Phi_+(s) = -\frac{s_2}{\mu}\left[\frac{-\mu/s_2}{s} + \frac{\mu+s_2}{s_2(s-s_2)}\right]$$

$$\Phi_+(s) = \frac{1}{s} - \frac{1+s_2/\mu}{s-s_2}$$

Inverting we obtain

$$W(y) = 1 - \left(1+\frac{s_2}{\mu}\right)e^{s_2 y} \quad y\geq 0$$

[Note: This $H_2/M/1$ system is a $G/M/1$ system, where it is easily seen that $\sigma = 1 + \frac{s_2}{\mu}$ or $-s_2 = \mu(1-\sigma)$.]

8.8.

PROBLEM 8.9.

Consider a G/G/1 system in which

$$A^*(s) = \frac{2}{(s+1)(s+2)}$$

$$B^*(s) = \frac{1}{s+1}$$

(a) Find the expression for $\Psi_+(s)/\Psi_-(s)$ and show the pole-zero plot in the s-plane.
(b) Use spectrum factorization to find $\Psi_+(s)$ and $\Psi_-(s)$.
(c) Find $\Phi_+(s)$.
(d) Find $W(y)$.
(e) Find the average waiting time W.
(f) We solved for $W(y)$ by the method of spectrum factorization. Can you describe another way to find $W(y)$?

SOLUTION

(a)

$$\frac{\Psi_+(s)}{\Psi_-(s)} = A^*(-s)B^*(s) - 1 = \frac{2}{(1-s)(2-s)} \frac{1}{s+1} - 1$$

$$= \frac{-s(s^2 - 2s - 1)}{(1-s)(2-s)(1+s)}$$

$$= \frac{-s[s - (1+\sqrt{2})][s - (1-\sqrt{2})]}{(1-s)(2-s)(1+s)} \qquad \blacksquare$$

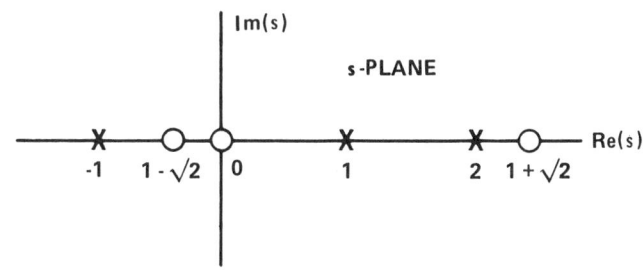

Zeroes (O) and poles (×) of $\Psi_+(s)/\Psi_-(s)$

8.9.

(b) Using (8.36) and (8.37) we find
$$\Psi_+(s) = \frac{s\,[s-(1-\sqrt{2})]}{s+1}$$

$$\Psi_-(s) = -\frac{(1-s)\,(2-s)}{s-(1+\sqrt{2})}$$

(pick $0 < D \leq 1$)

(c)
$$\Phi_+(s) = \frac{K}{\Psi_+(s)}$$

where $K = \lim_{s \to 0} \dfrac{\Psi_+(s)}{s}$. Thus $K = \sqrt{2}-1$ and so

$$\Phi_+(s) = (\sqrt{2}-1)\frac{s+1}{s\,[s-(1-\sqrt{2})]}$$

or

$$\Phi_+(s) = \frac{1}{s} - \frac{2-\sqrt{2}}{s-(1-\sqrt{2})}$$

(d) Inverting this last we obtain
$$W(y) = 1 - (2-\sqrt{2})\,e^{-(\sqrt{2}-1)y} \quad y \geq 0$$

(e) For W, we recognize that $W(y)$ is exponential in character. Indeed its density is
$$w(y) = (\sqrt{2}-1)\,u_0(y) + (2-\sqrt{2})\,(\sqrt{2}-1)\,e^{-(\sqrt{2}-1)y} \quad y \geq 0$$

By inspection, the mean wait W is
$$W = \frac{2-\sqrt{2}}{\sqrt{2}-1} = \sqrt{2}$$

(f) Noting that this is a G/M/1 system, we may use the techniques of Chapter 6 to solve for $W(y)$. Thus, first find σ satisfying $\sigma = A^*(\mu - \mu\sigma)$ (this gives $\sigma = 2-\sqrt{2}$) and then
$$W(y) = 1 - \sigma e^{-\mu(1-\sigma)y} \quad y \geq 0$$

by Eq. (6.30).

8.9.

PROBLEM 8.10.

Consider the system M/G/1. Using the spectral solution method for Lindley's integral equation, find
(a) $\Psi_+(s)$. {HINT: Interpret $[1 - B^*(s)]/s\bar{x}$.}
(b) $\Psi_-(s)$.
(c) $s\Phi_+(s)$.

SOLUTION

(a) For M/G/1, $A^*(s) = \dfrac{\lambda}{s+\lambda}$. Thus

$$\frac{\Psi_+(s)}{\Psi_-(s)} = A^*(-s)B^*(s) - 1$$

$$= \frac{\lambda}{\lambda - s}B^*(s) - 1$$

$$= \frac{s - \lambda + \lambda B^*(s)}{\lambda - s}$$

Factoring the zero at $s = 0$ gives

$$\frac{\Psi_+(s)}{\Psi_-(s)} = \frac{s\left[1 - \dfrac{\lambda\bar{x}[1 - B^*(s)]}{s\bar{x}}\right]}{\lambda - s}$$

$$= \frac{s[1 - \rho\hat{B}^*(s)]}{\lambda - s}$$

where we recall from Eq. (5.107) that

$$\hat{B}^*(s) \triangleq \frac{1 - B^*(s)}{s\bar{x}}$$

is the transform of the residual life pdf for the service time. Proceeding as on page 383 in Appendix II, we know that the density function of a non-negative random variable has a Laplace transform, say $H^*(s)$, which, for $\text{Re}(s) \geq 0$, is analytic and satisfies $|H^*(s)| \leq 1$. Applying this to $\hat{B}^*(s)$ we find

$$|\rho \hat{B}^*(s)| \leq \rho < 1 \text{ for } \text{Re}(s) \geq 0.$$

Thus $1 - \rho \hat{B}^*(s)$ has no zeroes (and no poles) in $\text{Re}(s) \geq 0$, and so this function must be part of $\Psi_+(s)$. Using (8.36) and (8.37) we have

$$\Psi_+(s) = s[1 - \rho \hat{B}^*(s)] = s - \lambda + \lambda B^*(s) \qquad \blacksquare$$

[Note that $\lim_{|s| \to \infty} \hat{B}^*(s) = 0$ for $\text{Re}(s) > 0$ so (8.37) holds.]

(b) Clearly,

$$\Psi_-(s) = \lambda - s \qquad \blacksquare$$

(pick $0 < D \leq \lambda$)

(c) $\Phi_+(s) = \dfrac{K}{\Psi_+(s)}$ where $K = \lim_{s \to 0} \dfrac{\Psi_+(s)}{s}$. Thus

$$K = \lim_{s \to 0} [1 - \rho \hat{B}^*(s)] = 1 - \rho$$

and

$$\Phi_+(s) = \dfrac{1-\rho}{s - \lambda + \lambda B^*(s)}$$

Finally

$$s\Phi_+(s) = \dfrac{s(1-\rho)}{s - \lambda + \lambda B^*(s)} = W^*(s)$$

This is Eq. (5.105), the P—K transform equation for the waiting-time distribution.

PROBLEM 8.11.

Consider the queue $E_q/E_r/1$.
(a) Show that

$$\dfrac{\Psi_+(s)}{\Psi_-(s)} = \dfrac{F(s)}{1 - F(s)}$$

where $F(s) = 1 - (1 - s/\lambda q)^q (1 + s/\mu r)^r$.

(b) For $\rho < 1$, show that $F(s)$ has one zero at the origin, zeroes $s_1, s_2, \ldots, s_r$ in $\text{Re}(s) < 0$, and zeroes $s_{r+1}, s_{r+2}, \ldots, s_{r+q-1}$ in $\text{Re}(s) > 0$.
(c) Express $\Psi_+(s)$ and $\Psi_-(s)$ in terms of s_i.
(d) Express $W^*(s)$ in terms of s_i ($i = 1, 2, \ldots, r+q-1$).

SOLUTION

(a) For $E_q/E_r/1$, we have $A^*(s) = \left(\dfrac{\lambda q}{s+\lambda q}\right)^q$ and $B^*(s) = \left(\dfrac{\mu r}{s+\mu r}\right)^r$. (Note: $\bar{t} = 1/\lambda$, $\bar{x} = 1/\mu$ and so $\rho = \lambda/\mu$).

$$\frac{\Psi_+(s)}{\Psi_-(s)} = A^*(-s)B^*(s) - 1 = \left(\frac{\lambda q}{\lambda q - s}\right)^q \left(\frac{\mu r}{\mu r + s}\right)^r - 1$$

$$= \frac{1}{\left(1 - \dfrac{s}{\lambda q}\right)^q \left(1 + \dfrac{s}{\mu r}\right)^r} - 1$$

$$= \frac{1 - \left(1 - \dfrac{s}{\lambda q}\right)^q \left(1 + \dfrac{s}{\mu r}\right)^r}{\left(1 - \dfrac{s}{\lambda q}\right)^q \left(1 + \dfrac{s}{\mu r}\right)^r}$$

Thus

$$\frac{\Psi_+(s)}{\Psi_-(s)} = \frac{F(s)}{1 - F(s)}$$

where

$$F(s) = 1 - \left(1 - \frac{s}{\lambda q}\right)^q \left(1 + \frac{s}{\mu r}\right)^r$$

(b) We assume $\rho < 1$ or $\lambda < \mu$. We wish to apply Rouche's theorem to $F(s)$. Thus we define the functions $f(s)$ and $g(s)$ as follows:

$$f(s) = -\left(1 - \frac{s}{\lambda q}\right)^q \left(1 + \frac{s}{\mu r}\right)^r \quad \text{and} \quad g(s) = 1.$$

Therefore $F(s) = f(s) + g(s)$. We first consider two circles: let K_1 be the circle of radius μr centered at the point $s \triangleq (x, y) = (-\mu r, 0)$ and let K_2 be the circle of radius λq centered at the point $s \triangleq (x, y) = (\lambda q, 0)$.

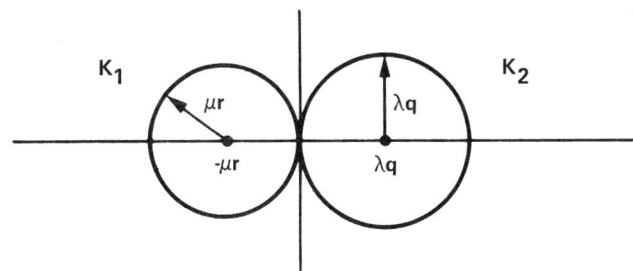

8.11.

Any point $s = (x,y)$ on K_1 must satisfy $(x+\mu r)^2 + y^2 = (\mu r)^2$. Thus we have $|1+s/\mu r|^2 = |(\mu r + x + jy)/\mu r|^2 = \dfrac{(x+\mu r)^2 + y^2}{(\mu r)^2} = 1$ on K_1. Hence outside of K_1, $|1+s/\mu r|^2 > 1$. Similarly $|1-s/\lambda q|^2 = 1$ on K_2 and $|1-s/\lambda q|^2 > 1$ outside of K_2. Thus $|f(s)| > 1 = |g(s)|$ for any point s which is outside one of the circles and on or outside the other. Note that $s = 0$ does not satisfy this condition, and to use Rouche's theorem we will have to detour around the origin. To this end, let $K_3(\epsilon)$ be a semicircle of radius $\epsilon > 0$ about the origin in the left-hand plane (i.e. $\mathrm{Re}(s) < 0$). On $K_3(\epsilon)$, $s = (x, y)$ must satisfy $x^2 + y^2 = \epsilon^2$, $x = \epsilon \cos\theta$, $y = \epsilon \sin\theta$, where $\dfrac{\pi}{2} < \theta < \dfrac{3\pi}{2}$ (i.e. $\cos\theta < 0$).

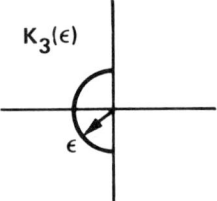

On $K_3(\epsilon)$ we have

$$|1-s/\lambda q|^2 = \frac{(\lambda q - x)^2 + y^2}{(\lambda q)^2} = \frac{(\lambda q)^2 - 2\lambda q x + \epsilon^2}{(\lambda q)^2}$$

Since $\epsilon > 0$, then we have

$$|1-s/\lambda q|^2 > \frac{(\lambda q)^2 - 2\lambda q x}{(\lambda q)^2} = 1 - \frac{2\epsilon \cos\theta}{\lambda q}$$

Similarly on $K_3(\epsilon)$

$$|1+s/\mu r|^2 > 1 + \frac{2\epsilon \cos\theta}{\mu r}$$

Thus, on $K_3(\epsilon)$, we have

$$|f(s)|^2 = \left| -\left(1 - \frac{s}{\lambda q}\right)^q \left(1 + \frac{s}{\mu r}\right)^r \right|^2$$

$$> \left(1 - \frac{2\epsilon \cos\theta}{\lambda q}\right)^q \left(1 + \frac{2\epsilon \cos\theta}{\mu r}\right)^r$$

$$= \left(1 - \frac{2\epsilon \cos\theta}{\lambda} + o(\epsilon)\right)\left(1 + \frac{2\epsilon \cos\theta}{\mu} + o(\epsilon)\right)$$

8.11.

and so

$$|f(s)|^2 > 1 - \frac{2\epsilon\cos\theta}{\lambda}(1-\rho) + o(\epsilon)$$

Since $\cos\theta < 0$ and $\rho < 1$ then $-\frac{2\epsilon\cos\theta}{\lambda}(1-\rho) > 0$. Thus from the definition of $o(\epsilon)$, there exists $\delta > 0$ such that for $0 < \epsilon < \delta$ $|f(s)| > 1 = |g(s)|$ on $K_3(\epsilon)$. Now we apply Rouche's theorem as follows: Let $C_1(\epsilon)$ be the closed contour consisting of part of K_1 plus the detour around the origin in $\mathrm{Re}(s) < 0$ ($0 < \epsilon < \delta$).

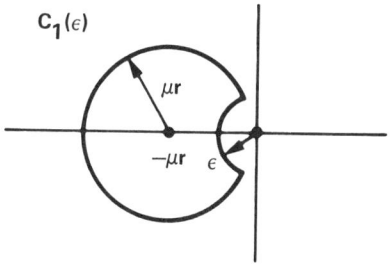

Then $|f(s)| > 1 = |g(s)|$ on $C_1(\epsilon)$. $f(s)$ has r roots inside $C_1(\epsilon)$, all at $s = (-\mu r, 0)$. Thus $F(s) = f(s) + g(s)$ has r roots inside $C_1(\epsilon)$. Letting $\epsilon \to 0$ we see that $F(s)$ has r zeroes $s_1, \ldots, s_r$ inside K_1, the circle of radius μr about $s = (-\mu r, 0)$. Next let $C_2(\epsilon)$ be the closed contour consisting of part of K_2 together with part of a circle about the origin of radius ϵ ($C_2(\epsilon)$ contains the origin).

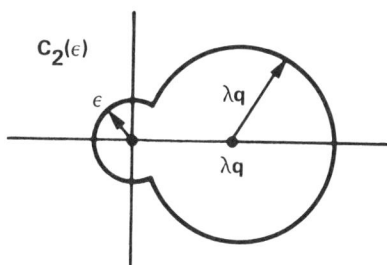

As noted above, for any point s on $C_2(\epsilon)$ outside of K_1 (which includes all points on $C_2(\epsilon)$ in $\mathrm{Re}(s) \geq 0$) we have $|f(s)| > 1 = |g(s)|$. For points s on $C_2(\epsilon)$ in $\mathrm{Re}(s) < 0$ we use the same argument as before. Thus $|f(s)| > 1 = |g(s)|$ on $C_2(\epsilon)$. $f(s)$ has q zeroes inside $C_2(\epsilon)$ (all at $s = (\lambda q, 0)$), so $F(s)$ has q zeroes inside $C_2(\epsilon)$. By examining $F(s)$ (recall that $\rho < 1$) we see that exactly one of these zeroes is at

$s = 0$. Letting $\epsilon \to 0$ we see that $F(s)$ has $q-1$ zeroes $s_{r+1}, \ldots, s_{r+q-1}$ inside the circle K_2. In summary: $F(s)$ has r zeroes $s_1, \ldots, s_r$ inside the circle of radius μr centered at $s = (-\mu r, 0)$, $F(s)$ has one zero at the origin, and $F(s)$ has $q-1$ zeroes $s_{r+1}, \ldots, s_{r+q-1}$ inside the circle of radius λq centered at $s = (\lambda q, 0)$. This (stronger) statement implies part (b).

(c) From part (b), we can write

$$F(s) = 1 - \left(1 - \frac{s}{\lambda q}\right)^q \left(1 + \frac{s}{\mu r}\right)^r$$

$$= -\left(\frac{-1}{\lambda q}\right)^q \left(\frac{1}{\mu r}\right)^r s(s - s_1) \cdots (s - s_{r+q-1})$$

So

$$\frac{\Psi_+(s)}{\Psi_-(s)} = \frac{F(s)}{1 - F(s)} = \frac{F(s)}{\left(1 - \frac{s}{\lambda q}\right)^q \left(1 + \frac{s}{\mu r}\right)^r}$$

$$= -\frac{s(s - s_1) \cdots (s - s_{r+q-1})}{(s - \lambda q)^q (s + \mu r)^r}$$

$\Psi_+(s)$ must absorb all the poles and zeroes in the left-hand plane, thus

$$\Psi_+(s) = \frac{s(s - s_1) \cdots (s - s_r)}{(s + \mu r)^r} \qquad \blacksquare$$

$$\Psi_-(s) = -\frac{(s - \lambda q)^q}{(s - s_{r+1}) \cdots (s - s_{r+q-1})} \qquad \blacksquare$$

or

$$\Psi_+(s) = \frac{s \prod_{i=1}^{r}(s - s_i)}{(s + \mu r)^r} \qquad \blacksquare$$

$$\Psi_-(s) = -\frac{(s - \lambda q)^q}{\prod_{i=r+1}^{r+q-1}(s - s_i)} \qquad \blacksquare$$

8.11.

(d) $\Phi_+(s) = \dfrac{K}{\Psi_+(s)}$ where $K = \lim\limits_{s \to 0} \dfrac{\Psi_+(s)}{s}$. Thus

$$K = \dfrac{\prod\limits_{i=1}^{r}(-s_i)}{(\mu r)^r}$$

and so

$$\Phi_+(s) = \dfrac{\prod\limits_{i=1}^{r}(-s_i)}{(\mu r)^r} \cdot \dfrac{(s+\mu r)^r}{s\prod\limits_{i=1}^{r}(s-s_i)}$$

$$W^*(s) = s\,\Phi_+(s) = \dfrac{\left(1 + \dfrac{s}{\mu r}\right)^r}{\prod\limits_{i=1}^{r}\left(1 - \dfrac{s}{s_i}\right)} \qquad \blacksquare$$

PROBLEM 8.12.

Show that Eq. (8.71) is equivalent to Eq. (6.30).

SOLUTION

Eq. (8.71) is

$$W(y) = 1 - \left(1 - \dfrac{s_1}{\mu}\right)e^{-s_1 y} \quad y \geqslant 0$$

where $-s_1$ is the only zero of $s + \mu - \mu A^*(-s)$ satisfying $\mathrm{Re}(s) < 0$. Thus $-s_1 + \mu = \mu A^*(s_1)$. Define $\sigma = 1 - \dfrac{s_1}{\mu}$. Therefore $s_1 = \mu - \mu\sigma$. So we have $\mu\sigma = \mu A^*(\mu - \mu\sigma)$ or $\sigma = A^*(\mu - \mu\sigma)$. Eq. (8.71) becomes

$$W(y) = 1 - \sigma e^{-\mu(1-\sigma)y} \quad y \geqslant 0$$

which is Eq. (6.30).
[Note: by Eq. (8.69), $\sigma = 1 - W(0^+)$. Thus $0 < \sigma < 1$.]

PROBLEM 8.13.

Consider a D/D/1 queue with $\rho < 1$. Assume $w_0 = 4\bar{t}(1-\rho)$.
(a) Calculate $w_n(y)$ using the procedure defined by Eq. (8.78) for $n = 0, 1, 2, \ldots$.
(b) Show that the known solution for
$$w(y) = \lim_{n \to \infty} w_n(y)$$
satisfies Eq. (8.79).

SOLUTION

(a) Eq. (8.78) states
$$w_{n+1}(y) = \pi(c(y) \circledast w_n(y)) \quad n = 1, 2, \ldots$$
For D/D/1 we know
$$a(y) = u_0(y - \bar{t})$$
$$b(y) = u_0(y - \bar{x})$$
Then
$$c(y) = a(-y) \circledast b(y) \quad \text{(Eq. (8.18))}$$
So
$$c(y) = u_0(y + \bar{t} - \bar{x}) = u_0(y + \bar{t}(1-\rho))$$
Thus
$$c(y) \circledast w_n(y) = \int_{-\infty}^{y} w_n(y-t) c(t)\, dt$$
$$= w_n(y + \bar{t}(1-\rho))$$
So
$$w_{n+1}(y) = \pi(w_n(y + \bar{t}(1-\rho)))$$
Using this relationship w_n may be calculated for $n = 0, 1, 2, \ldots$. By assumption
$$w_0 = u_0(y - 4\bar{t}(1-\rho))$$ ■

Also
$$w_1(y) = \pi(w_0(y + \bar{t}(1-\rho))) = \pi(u_0(y - 3\bar{t}(1-\rho)))$$

8.13.

205

Since the sweep operator π has no "negative" probability to sweep up, we find

$$w_1(y) = u_0(y - 3\bar{t}(1-\rho))$$ ∎

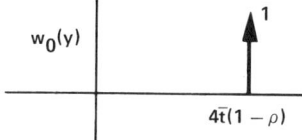 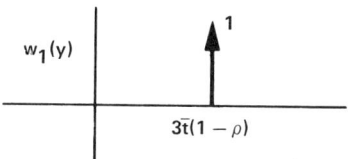

$$w_2(y) = \pi(w_1(y + \bar{t}(1-\rho))) = \pi(u_0(y - 2\bar{t}(1-\rho)))$$

$$w_2(y) = u_0(y - 2\bar{t}(1-\rho))$$ ∎

Similarly

$$w_3(y) = u_0(y - \bar{t}(1-\rho))$$ ∎

and

$$w_4(y) = u_0(y)$$ ∎

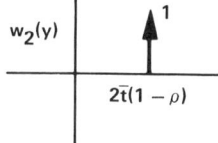 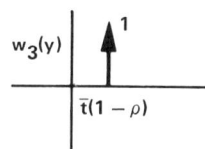 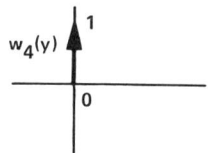

In fact,

$$w_n(y) = u_0(y) \text{ for } n \geq 4$$ ∎

(due to sweeping)

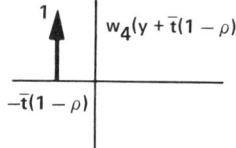  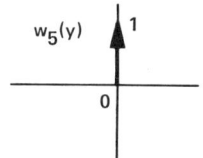

8.13.

(b)

$$w(y) = \lim_{n\to\infty} w_n(y) = u_0(y) \qquad \blacksquare$$

$$c(y) \circledast w(y) = \int_{-\infty}^{y} w(y-t)c(t)\,dt$$

$$= \int_{-\infty}^{y} w(y-t)u_0(t + \bar{t}(1-\rho))\,dt$$

$$= w(y + \bar{t}(1-\rho)) = u_0(y + \bar{t}(1-\rho))$$

So

$$\pi(c(y) \circledast w(y)) = \pi(u_0(y + \bar{t}(1-\rho)))$$

$$= u_0(y)$$

(the unit impulse at $y = -\bar{t}(1-\rho)$ is swept up to the origin)

$$\pi(c(y) \circledast w(y)) = w(y)$$

and Eq. (8.79) is satisfied.

PROBLEM 8.14.

Consider an M/M/1 queue with $\rho < 1$. Assume $w_0 = 0$.
(a) Calculate $w_1(y)$ using the procedure defined by Eq. (8.78).
(b) Repeat for $w_2(y)$.
(c) Show that our known solution for

$$w(y) = \lim_{n\to\infty} w_n(y)$$

satisfies Eq. (8.79).
(d) Compare $w_2(y)$ with $w(y)$.

SOLUTION

(a) Differentiating the result for $C(y)$ in Exercise (8.2), we find for M/M/1

$$c(y) = \begin{cases} \dfrac{\lambda\mu}{\lambda+\mu} e^{-\mu y} & y \geq 0 \\[2mm] \dfrac{\lambda\mu}{\lambda+\mu} e^{\lambda y} & y < 0 \end{cases}$$

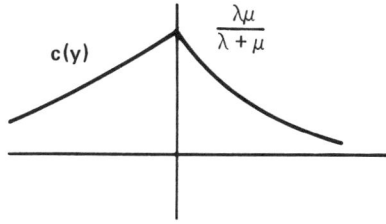

$w_0 = 0$ implies $w_0(y) = u_0(y)$ $(y \geq 0)$. By Eq. (8.78) we have

$$w_1(y) = \pi(c(y) \circledast w_0(y))$$

$$w_1(y) = \pi(c(y) \circledast u_0(y)) = \pi(c(y))$$

Thus

$$w_1(y) = u_0(y) \int_{-\infty}^{0} \frac{\lambda\mu}{\lambda+\mu} e^{\lambda y}\, dy + \frac{\lambda\mu}{\lambda+\mu} e^{-\mu y}$$

$$w_1(y) = \frac{\mu}{\lambda+\mu} u_0(y) + \frac{\lambda\mu}{\lambda+\mu} e^{-\mu y} \quad y \geq 0 \qquad \blacksquare$$

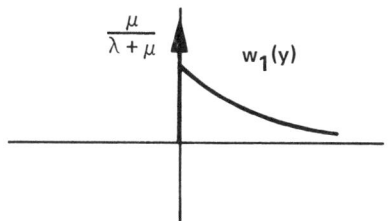

(b)

$$w_2(y) = \pi(c(y) \circledast w_1(y))$$

We first determine $c(y) \circledast w_1(y)$.

$$c(y) \circledast w_1(y) = \int_{-\infty}^{\infty} c(y-t) w_1(t)\, dt = \int_{0}^{\infty} c(y-t) w_1(t)\, dt$$

Case (1): $y \leq 0$
Here $y - t \leq 0$ for all $0 \leq t < \infty$

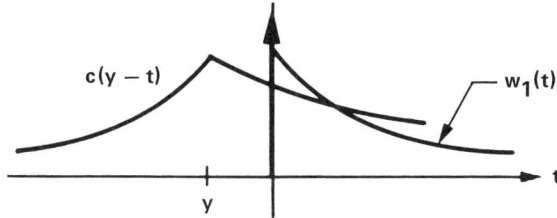

8.14.

$$\therefore c(y) \circledast w_1(y) = \int_t^\infty \frac{\lambda\mu}{\lambda+\mu} e^{\lambda(y-t)} \left[\frac{\mu}{\lambda+\mu} u_0(t) + \frac{\lambda\mu}{\lambda+\mu} e^{-\mu t}\right] dt$$

$$= \frac{\lambda\mu^2}{(\lambda+\mu)^2} e^{\lambda y} + \left(\frac{\lambda\mu}{\lambda+\mu}\right)^2 e^{\lambda y} \int_0^\infty e^{-(\lambda+\mu)t} dt$$

$$= \frac{\lambda\mu^2}{(\lambda+\mu)^2} e^{\lambda y} \left(1 + \frac{\lambda}{\lambda+\mu}\right)$$

Case (2): $y > 0$

We consider the intervals $y-t \geq 0$ and $y-t \leq 0$ i.e. $0 \leq t \leq y$ and $y \leq t < \infty$

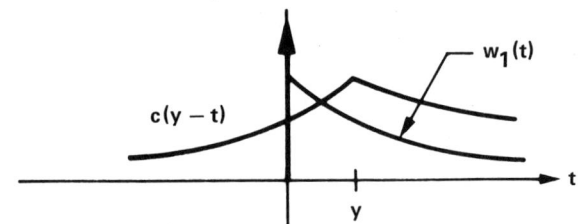

$$c(y) \circledast w_1(y) = \int_0^y \frac{\lambda\mu}{\lambda+\mu} e^{-\mu(y-t)} \left[\frac{\mu}{\lambda+\mu} u_0(t) + \frac{\lambda\mu}{\lambda+\mu} e^{-\mu t}\right] dt$$

$$+ \int_y^\infty \frac{\lambda\mu}{\lambda+\mu} e^{\lambda(y-t)} \left[\frac{\mu}{\lambda+\mu} u_0(t) + \frac{\lambda\mu}{\lambda+\mu} e^{-\mu t}\right] dt$$

or

$$c(y) \circledast w_1(y) = \frac{\lambda\mu^2}{(\lambda+\mu)^2} e^{-\mu y} + \left(\frac{\lambda\mu}{\lambda+\mu}\right)^2 e^{-\mu y} \int_0^y dt$$

$$+ \left(\frac{\lambda\mu}{\lambda+\mu}\right)^2 e^{\lambda y} \int_y^\infty e^{-(\lambda+\mu)t} dt$$

Thus

$$c(y) \circledast w_1(y) = \frac{\lambda\mu^2}{(\lambda+\mu)^2} e^{-\mu y}(1+\lambda y) + \frac{(\lambda\mu)^2}{(\lambda+\mu)^3} e^{\lambda y}[e^{-(\lambda+\mu)y}]$$

$$= \frac{\lambda\mu^2}{(\lambda+\mu)^2} e^{-\mu y}\left(1 + \lambda y + \frac{\lambda}{\lambda+\mu}\right)$$

8.14.

To find the probability in the negative half line we calculate

$$\int_{-\infty}^{0} c(y) \circledast w_1(y)\, dy = \int_{-\infty}^{0} \frac{\lambda\mu^2}{(\lambda+\mu)^2} e^{\lambda y}\left[1 + \frac{\lambda}{\lambda+\mu}\right] dy$$

$$= \frac{\lambda\mu^2}{(\lambda+\mu)^2}\left[1 + \frac{\lambda}{\lambda+\mu}\right] \frac{e^{\lambda y}}{\lambda}\bigg|_{-\infty}^{0}$$

$$= \left(\frac{\mu}{\lambda+\mu}\right)^2 \left[1 + \frac{\lambda}{\lambda+\mu}\right] = \frac{\mu^2(2\lambda+\mu)}{(\lambda+\mu)^3}$$

Thus

$$w_2(y) = \pi(c(y) \circledast w_1(y))$$

$$w_2(y) = \frac{\mu^2(2\lambda+\mu)}{(\lambda+\mu)^3} u_0(y) + \frac{\lambda\mu^2}{(\lambda+\mu)^2} e^{-\mu y}\left[1 + \lambda y + \frac{\lambda}{\lambda+\mu}\right] \quad y \geq 0 \quad \blacksquare$$

(c) The known solution for M/M/1 as given in Eq. (5.122) is

$$w(y) = (1-\rho)u_0(y) + \lambda(1-\rho)e^{-\mu(1-\rho)y} \quad y \geq 0$$

We first find $c(y) \circledast w(y)$.

$$c(y) \circledast w(y) = \int_{-\infty}^{\infty} c(y-t) w(t)\, dt = \int_{0}^{\infty} c(y-t) w(t)\, dt$$

For $y \leq 0$ then $y - t \leq 0$, so

$$c(y) \circledast w(y) = \int_{0}^{\infty} \frac{\lambda\mu}{\lambda+\mu} e^{\lambda(y-t)}\left[(1-\rho)u_0(t) + \lambda(1-\rho)e^{-\mu(1-\rho)t}\right] dt$$

$$= \frac{\lambda\mu}{\lambda+\mu} e^{\lambda y}(1-\rho) + \frac{\lambda\mu}{\lambda+\mu} e^{\lambda y} \lambda(1-\rho) \int_{0}^{\infty} e^{-\mu t}\, dt$$

$$= \frac{\lambda\mu}{\lambda+\mu} e^{\lambda y}(1-\rho)\left[1 + \frac{\lambda}{\mu}\right] = \lambda(1-\rho)e^{\lambda y}$$

Thus the probability π will sweep up is

$$\int_{-\infty}^{0} c(y) \circledast w(y)\, dy = \int_{-\infty}^{0} \lambda(1-\rho)e^{\lambda y}\, dy = 1 - \rho$$

For $y \geq 0$ we consider $0 \leq t \leq y$ and $y \leq t \leq \infty$. Then

$$c(y) \circledast w(y) = \int_{0}^{y} \frac{\lambda\mu}{\lambda+\mu} e^{-\mu(y-t)}\left[(1-\rho)u_0(t) + \lambda(1-\rho)e^{-\mu(1-\rho)t}\right] dt$$

$$+ \int_{y}^{\infty} \frac{\lambda\mu}{\lambda+\mu} e^{\lambda(y-t)}\left[(1-\rho)u_0(t) + \lambda(1-\rho)e^{-\mu(1-\rho)t}\right] dt$$

8.14.

or

$$c(y) \circledast w(y) = \frac{\lambda\mu}{\lambda+\mu}e^{-\mu y}(1-\rho) + \frac{\lambda\mu}{\lambda+\mu}e^{-\mu y}\lambda(1-\rho)\int_0^y e^{\lambda t}\,dt$$

$$+ \frac{\lambda\mu}{\lambda+\mu}e^{\lambda y}\lambda(1-\rho)\int_y^\infty e^{-\mu t}\,dt$$

$$= \frac{\lambda\mu}{\lambda+\mu}e^{-\mu y}(1-\rho)e^{\lambda y} + \frac{\lambda\mu}{\lambda+\mu}e^{\lambda y}\lambda(1-\rho)\frac{e^{-\mu y}}{\mu}$$

$$= \frac{\lambda\mu}{\lambda+\mu}e^{-(\mu-\lambda)y}(1-\rho)\left[1 + \frac{\lambda}{\mu}\right]$$

$$= \lambda(1-\rho)e^{-\mu(1-\rho)y}$$

Thus
$$\pi(c(y) \circledast w(y)) = (1-\rho)u_0(y) + \lambda(1-\rho)e^{-\mu(1-\rho)y} \quad y \geq 0$$
and
$$w(y) = \pi(c(y) \circledast w(y))$$

Hence $w(y)$ satisfies Eq. (8.79).

(d)
$$w(y) = (1-\rho)u_0(y) + \lambda(1-\rho)e^{-\mu(1-\rho)y} \quad y \geq 0$$

$$w_2(y) = \frac{\mu^2(2\lambda+\mu)}{(\lambda+\mu)^3}u_0(y) + \frac{\lambda\mu^2}{(\lambda+\mu)^2}e^{-\mu y}\left[1 + \lambda y + \frac{\lambda}{\lambda+\mu}\right] \quad y \geq 0$$

$$P[w_2 = 0] = \frac{\mu^2(2\lambda+\mu)}{(\lambda+\mu)^3} = \frac{1+2\rho}{(1+\rho)^3}$$

$$> \frac{(1+2\rho+\rho^2)(1-\rho^2)}{(1+\rho)^3} = 1-\rho$$

or
$$P[w_2 = 0] > P[\tilde{w} = 0]$$

So $w_2(y)$ has a larger impulse at the origin than $w(y)$. Also we note that $w(y)$ decays as $e^{-\mu(1-\rho)y} = e^{-\mu y}e^{\lambda y}$, whereas $w_2(y)$ decays as $e^{-\mu y}\left[1 + \lambda y + \frac{\lambda}{\lambda+\mu}\right]$. Thus $w_2(y)$ decays more rapidly.

8.14.

PROBLEM 8.15.

By first cubing Eq. (8.91) and then forming expectations, express $\sigma_{\tilde{w}}^2$ (the variance of the waiting time) in terms of the first three moments of $\tilde{t}$, $\tilde{x}$, and I.

SOLUTION

Cubing Eq. (8.91) ($w_{n+1} - y_n = w_n + u_n$) gives

$$w_{n+1}^3 - 3w_{n+1}^2 y_n + 3w_{n+1} y_n^2 - y_n^3 = w_n^3 + 3w_n^2 u_n + 3 w_n u_n^2 + u_n^3$$

Recalling that $w_{n+1} y_n = 0$, taking expectations and letting $n \to \infty$, we get

$$\overline{w^3} - \overline{y^3} = \overline{w^3} + 3\overline{w^2 u} + 3\overline{w u^2} + \overline{u^3}$$

Since $\tilde{w}$ and $\tilde{u}$ are independent we find

$$\overline{w^2} = -\frac{\overline{y^3} + \overline{u^3}}{3\overline{u}} - \frac{\overline{w}\,\overline{u^2}}{\overline{u}}$$

Thus

$$\sigma_w^2 = \overline{w^2} - (\overline{w})^2 = -\frac{\overline{y^3} + \overline{u^3}}{3\overline{u}} - \frac{\overline{w}\,\overline{u^2}}{\overline{u}} - (\overline{w})^2$$

Using Eq. (8.94) which states $\overline{w} = -\dfrac{\overline{u^2}}{2\overline{u}} - \dfrac{\overline{y^2}}{2\overline{y}}$ we get

$$\sigma_w^2 = -\frac{\overline{y^3} + \overline{u^3}}{3\overline{u}} + \frac{\overline{u^2}}{\overline{u}}\left(\frac{\overline{u^2}}{2\overline{u}} + \frac{\overline{y^2}}{2\overline{y}}\right) - \left(\frac{\overline{u^2}}{2\overline{u}} + \frac{\overline{y^2}}{2\overline{y}}\right)^2$$

$$= -\frac{\overline{y^3} + \overline{u^3}}{3\overline{u}} + \left(\frac{\overline{u^2}}{2\overline{u}} + \frac{\overline{y^2}}{2\overline{y}}\right)\left(\frac{\overline{u^2}}{2\overline{u}} - \frac{\overline{y^2}}{2\overline{y}}\right)$$

Eq. (8.92) gives $\overline{y} = -\overline{u}$, and Eq. (8.100) gives $\overline{y^k} = a_0 \overline{I^k}$. Thus

$$\sigma_w^2 = \frac{\overline{I^3}}{3\overline{I}} - \frac{\overline{u^3}}{3\overline{u}} + \left(\frac{\overline{u^2}}{2\overline{u}} + \frac{\overline{I^2}}{2\overline{I}}\right)\left(\frac{\overline{u^2}}{2\overline{u}} - \frac{\overline{I^2}}{2\overline{I}}\right)$$

Noting that $\tilde{u} = \tilde{x} - \tilde{t}$ we finally obtain

$$\sigma_w^2 = -\frac{\overline{x^3} - 3\overline{x^2 t} + 3\overline{x\, t^2} - \overline{t^3}}{3(\overline{x} - \overline{t})} + \frac{1}{4}\left[\frac{\overline{x^2} - 2\overline{x\,t} + \overline{t^2}}{\overline{x} - \overline{t}}\right]^2 + \frac{\overline{I^3}}{3\overline{I}} - \left(\frac{\overline{I^2}}{2\overline{I}}\right)^2$$

8.15.

or

$$\sigma_w^2 = \frac{\overline{x^3}-\overline{t^3}}{3\bar{t}(1-\rho)} + \frac{\rho\overline{t^2}-\overline{x^2}}{1-\rho} + \left[\frac{\sigma_a^2+\sigma_b^2+[\bar{t}(1-\rho)]^2}{2\bar{t}(1-\rho)}\right]^2 + \frac{\overline{t^3}}{3\bar{t}} - \left(\frac{\overline{t^2}}{2\bar{t}}\right)^2 \blacksquare$$

PROBLEM 8.16.

Show that $P[\tilde{w}=0] = 1-\sigma$ from Eq. (8.117) by finding the constant term in a power-series expansion of $W^*(s)$.

SOLUTION

Eq. (8.117) gives $W^*(s) = \dfrac{1-\sigma}{1-\sigma \hat{I}^*(s)}$. So

$$W^*(s) = (1-\sigma)\sum_{j=0}^{\infty} [\sigma \hat{I}^*(s)]^j$$

Inverting (see for example Eq. (5.109) and Eq. (5.111)) we have

$$w(y) = (1-\sigma)\sum_{j=0}^{\infty} \sigma^j \hat{i}_{(j)}(y)$$

where the random variable $\hat{I}_k$ has the density function $\hat{i}(y)$ with transform $\hat{I}^*(s)$. Thus

$$w(y) = (1-\sigma)u_0(y) + (1-\sigma)\sum_{j=1}^{\infty} \sigma^j \hat{i}_{(j)}(y)$$

In order to find $P[\tilde{w}=0]$, we must determine if $\hat{i}_{(j)}(y)$ contains an impulse at the origin ($y=0$) for each $j=1,2,\ldots$ From the definition of $\hat{I}_k$ (see Eq. (8.114)), we see that $P[\hat{I}_k=0]=0$, and so $\hat{i}_{(j)}(y)$ cannot contain an impulse at the origin. (Also note $\hat{I}_k$ is the idle period in a "dual queue", and so it must be that $P[\hat{I}_k > 0] = 1$.) Finally

$$P[\tilde{w}=0] = \int_{0^-}^{0^+} w(y)\,dy = 1-\sigma + (1-\sigma)\sum_{j=1}^{\infty} \sigma^j \int_{0^-}^{0^+} \hat{i}_{(j)}(y)\,dy$$

or

$$P[\tilde{w}=0] = 1-\sigma$$

PROBLEM 8.17.†

Consider a G/G/1 system.
(a) Express $\hat{I}^*(s)$ in terms of the transform of the pdf of idle time in the given system.
(b) Using (a) find $\hat{I}^*(s)$ when the original system is the ordinary M/M/1.
(c) Consider a G/M/1 queue.
   (i) For $\rho < 1$ (stable queue), use Eq. (8.106) to show directly that
   $$I^*(s) = \frac{A^*(s) - \sigma}{-\frac{s}{\mu} + 1 - \sigma}$$
   (ii) For $\rho > 1$ (unstable queue), use (a) to show that
   $$I^*(s) = \frac{1 - A^*(s)}{\bar{s}t}$$
   which gives Eq. (8.121) as promised.
(d) Since either the original or the dual queue must be unstable (except for D/D/1), discuss the existence of the transform of the idle-time pdf for the unstable queue.

SOLUTION

(a) Eq. (8.106) and Eq. (8.117) give expressions for $W^*(s)$. Equating these two yields
$$\frac{a_0[1 - I^*(-s)]}{1 - C^*(s)} = W^*(s) = \frac{1 - \sigma}{1 - \sigma \hat{I}^*(s)}$$
Solving for $\hat{I}^*(s)$ we find
$$\hat{I}^*(s) = \frac{1}{\sigma} - \frac{(1-\sigma)[1 - C^*(s)]}{\sigma a_0[1 - I^*(-s)]} \quad \blacksquare$$

Here $1 - \sigma = P[\tilde{w} = 0]$ and $a_0 = P[\tilde{y} > 0]$. Note that $1 - \sigma = a_0$ if there is zero probability that an arrival and a departure occur at the same instant of time (for example, when either the interarrival or service distribution is continuous). In this latter case we obtain the simplified expression
$$\hat{I}^*(s) = \frac{1}{\sigma}\left[1 - \frac{1 - C^*(s)}{1 - I^*(-s)}\right] \quad \blacksquare$$

(b) For M/M/1, $1-\sigma = a_0$ as noted above. Also, $\sigma = \rho$, $I^*(-s) = \dfrac{\lambda}{\lambda - s}$, and $C^*(s) = A^*(-s)B^*(s) = \dfrac{\lambda\mu}{(\mu+s)(\lambda-s)}$. Using part (a) we find

$$\hat{I}^*(s) = \frac{1}{\rho}\left[1 - \frac{1 - \dfrac{\lambda\mu}{(\mu+s)(\lambda-s)}}{1 - \dfrac{\lambda}{\lambda-s}}\right]$$

$$= \frac{1}{\rho}\left[1 - \frac{(\mu+s)(\lambda-s) - \lambda\mu}{(\mu+s)(-s)}\right]$$

$$= \frac{1}{\rho}\left[\frac{(\mu+s)(-\lambda) + \lambda\mu}{(\mu+s)(-s)}\right] = \frac{1}{\rho}\left(\frac{\lambda}{s+\mu}\right)$$

or

$$\hat{I}^*(s) = \frac{\mu}{s+\mu} \qquad \blacksquare$$

Thus the (unstable) dual queue has an exponential idle period distribution, which agrees with what we know since it is also M/M/1.

(c) **(i)** For a stable G/M/1 queue, from Eq. (6.30) we get

$$W^*(s) = \frac{(1-\sigma)(s+\mu)}{s+\mu(1-\sigma)}.$$

Also $a_0 = 1 - \sigma$ and $C^*(s) = A^*(-s)\dfrac{\mu}{s+\mu}$. Thus Eq. (8.106) gives

$$\frac{(1-\sigma)(s+\mu)}{s+\mu(1-\sigma)} = \frac{(1-\sigma)[1 - I^*(-s)]}{1 - A^*(-s)\dfrac{\mu}{s+\mu}}$$

or

$$1 - I^*(-s) = \frac{s+\mu - \mu A^*(-s)}{s+\mu - \mu\sigma}$$

$$I^*(-s) = \frac{\mu A^*(-s) - \mu\sigma}{s+\mu - \mu\sigma}$$

8.17.

and so
$$I^*(s) = \frac{A^*(s) - \sigma}{-\frac{s}{\mu} + 1 - \sigma}$$

(ii) Here we must exercise caution since Eq. (8.117), and thus the result of part (a), requires that the original queue be stable (which does not hold for the G/M/1 system we are now considering). Thus we regard our unstable G/M/1 queue with parameters $A^*(s), \mu$ as the dual of a stable M/G/1 queue with arrival rate μ and service time transform $A^*(s)$. For this stable M/G/1 system, $\sigma = \rho = \mu\bar{t}$, $I^*(s) = \frac{\mu}{s + \mu}$ and $C^*(s) = \frac{\mu}{\mu - s} A^*(s)$ (also note that $1 - \sigma = a_0$). We now use the result of part (a) to solve for $\hat{I}^*(s)$ which is, in fact, the idle time transform for our unstable G/M/1 queue. Thus

$$\hat{I}^*(s) = \frac{1}{\rho}\left[1 - \frac{1 - \frac{\mu}{\mu - s}A^*(s)}{1 - \frac{\mu}{\mu - s}}\right]$$

$$= \frac{1}{\rho}\left[1 - \frac{\mu - s - \mu A^*(s)}{-s}\right]$$

$$= \frac{1}{\mu\bar{t}}\left[\frac{-\mu[1 - A^*(s)]}{-s}\right]$$

or

$$\hat{I}^*(s) = \frac{1 - A^*(s)}{s\bar{t}}$$

which is the desired result.

(d) Although an idle period may never occur in the unstable queue, those that do occur will have durations properly described by the transforms discussed in the text.

8.17.

ERRATA SHEET

for

QUEUEING SYSTEMS
VOLUME I: THEORY

PAGE	LINE	ORIGINAL TEXT	CORRECTED TEXT
ix	21	agruments	arguments
x	Figure		
x	−5	[KLEI 74]	[KLEI 75]
xi	16	Chapter 5	Chapters 5 and 6
15	Fig. 2.2	Servicer	Server
17	*+1	number, $\bar{N}$, in	number in
17	*+2	This latter quantity is simply the arrival rate λ times his average time in system, T.	This latter quantity may be approximated by the arrival rate λ times his average time in system T, when T is large compared to the interarrival time.
27	4	we have (for $i_1 < i_2 < \ldots < i_n$) that	we have that
33	2	$\pi^{(n-1)}$	$\boldsymbol{\pi}^{(n-1)}$
46	1	(See Errata Sheet Appendix)	
50	1	result is*	result, assuming the values of $\mathbf{Q}(t)$ commute for different values of t, is*
67	−12	$P[\tilde{t} > t]$	$P[\tilde{t} > t_0]$
78	−4	**44**, 1−89	**44**, 1−189
79	Ex. 2.2	to the substream i.	to the ith substream.
80	Ex. 2.6	$P =$	$\mathbf{P} =$
83	Ex. 2.15	Consider a . . . $\mu_k = k\mu$.	Consider a . . . $\mu_k = k\mu$.
85	Ex. 2.20(e)	Using the fact	For $k \geqslant i$, using the fact
85	Ex. 2.20(e)	(See Errata Sheet Appendix)	
85	Ex. 2.20(f)	property 4	property 5
85	Ex. 2.20(f)	shown in part (e).	shown in part (e), thus establishing the transient solution for $k \geqslant i$.
85	Ex. 2.20(g)	(See Errata Sheet Appendix)	

218

PAGE	LINE	ORIGINAL TEXT	CORRECTED TEXT
85	Ex. 2.21	$\Phi_X(u)$	$\phi_X(u)$
87	7	$\pi = \pi \mathbf{P}$	$\pi \mathbf{Q} = 0$
93	−9	Eqs. (3.11) and (3.12).	Eqs. (3.11) and (3.12) for irreducible homogeneous Markov chains.
93‡	Eq. (3.18)	$\sum_{k=0}^{\infty} \prod_{i=0}^{k-1} \frac{\lambda_i}{\mu_{i+1}}$	$\sum_{k=1}^{\infty} \prod_{i=0}^{k-1} \frac{\lambda_i}{\mu_{i+1}}$
93‡	Eq. (3.19)	$\sum_{k=0}^{\infty} \left(1 / \left(\lambda_k \prod_{i=0}^{k-1} \frac{\lambda_i}{\mu_{i+1}} \right) \right)$	$\sum_{k=1}^{\infty} \prod_{i=1}^{k} \frac{\mu_i}{\lambda_i}$
94	2	(See Errata Sheet Appendix)	
94	8	(See Errata Sheet Appendix)	
94	11	(See Errata Sheet Appendix)	
95	Fig. 3.1	[diagram: node $k+1$ with arrows]	[diagram: node $k+1$ with arrows]
95‡	−8	$S_1 = \sum_{k=0}^{\infty} \frac{p_k}{p_0} = \sum_{k=0}^{\infty} \left(\frac{\lambda}{\mu}\right)^k < \infty$	$S_1 = \sum_{k=1}^{\infty} \prod_{i=0}^{k-1} \frac{\lambda_i}{\mu_{i+1}} = \sum_{k=1}^{\infty} \left(\frac{\lambda}{\mu}\right)^k < \infty$
95‡	−5	$S_2 = \sum_{k=0}^{\infty} \frac{1}{\lambda(p_k/p_0)} = \sum_{k=0}^{\infty} \frac{1}{\lambda}\left(\frac{\mu}{\lambda}\right)^k = \infty$	$S_2 = \sum_{k=1}^{\infty} \prod_{i=1}^{k} \frac{\mu_i}{\lambda_i} = \sum_{k=1}^{\infty} \left(\frac{\mu}{\lambda}\right)^k = \infty$
109	Fig. 3.12	$(M-m+1)\lambda$	$(M-m+2)\lambda$
109	Fig. 3.12	$(M-m+2)\lambda$	$(M-m+1)\lambda$
111	Ex. 3.6	$k = K_1, K_1+1, \ldots K_2$	$k = K_1, K_1+1, \ldots, K_2$
112	Ex. 3.9(b)	$C\left(m, \frac{\lambda}{\mu}\right) = \dfrac{B\left(m, \frac{\lambda}{\mu}\right)}{1 - \frac{\lambda}{\mu}\left[1 - B\left(m, \frac{\lambda}{\mu}\right)\right]}$	$C\left(m, \frac{\lambda}{\mu}\right) = \dfrac{B\left(m, \frac{\lambda}{\mu}\right)}{1 - \frac{\lambda}{m\mu}\left[1 - B\left(m, \frac{\lambda}{\mu}\right)\right]}$
112	Ex. 3.9(c)	$B\left(m+1, \frac{\lambda}{\mu}\right) = \dfrac{\frac{\mu}{\lambda} B\left(m, \frac{\lambda}{\mu}\right)}{m+1+\frac{\lambda}{\mu} B\left(m, \frac{\lambda}{\mu}\right)}$	$B\left(m+1, \frac{\lambda}{\mu}\right) = \dfrac{\frac{\lambda}{\mu} B\left(m, \frac{\lambda}{\mu}\right)}{m+1+\frac{\lambda}{\mu} B\left(m, \frac{\lambda}{\mu}\right)}$
114	Ex. 3.14	$M/M/1/K$	$M/M/1/K$
117	Fig. 4.1	$\frac{\lambda}{2}$	$\frac{\lambda}{2}$
118	3	$\bar{x} < \bar{t}$	$\bar{x} \leqslant \bar{t}$
127	header	$M/E_r/1$	$M/E_r/1$
127	Fig. 4.6	[diagram: node r]	[diagram: node r with λ]
127	Fig. 4.6	$M/E_r/1$	$M/E_r/1$

---

‡These simplified expressions give equivalent convergent conditions for state classification on page 94.

PAGE	LINE	ORIGINAL TEXT	CORRECTED TEXT
129	header	$M/E_r/1$	$M/E_r/1$
129	Eq. (4.29)	$j = 1, 2, \ldots, r$	$j = 1, 2, \ldots$
130	5	$E_r/M/1$	$E_r/M/1$
131	header	$E_r/M/1$	$E_r/M/1$
131	Fig. 4.7	$E_r/M/1$	$E_r/M/1$
133	header	$E_r/M/1$	$E_r/M/1$
136	9	$p_0 = 1 - \rho.$	$p_0 = 1 - \rho$ where $\rho = \lambda G'(1)/\mu.$
137	-9	available.	available; however, any customer who arrives to find less than r customers in service will immediately join them.
145	Eq. (4.63)	$B^*(s) = \sum_{j=1}^{R} \alpha_i \prod_{i=1}^{r_i} \left(\frac{\mu_{ij}}{s + \mu_{ij}} \right)$	$B^*(s) = \sum_{i=1}^{R} \alpha_i \prod_{j=1}^{r_i} \left(\frac{\mu_{ij}}{s + \mu_{ij}} \right)$
146	2	$R_e(s)$	$\mathrm{Re}(s)$
155	*+6	(r_{0i})	(r_{0j})
156	18	$\pi = \pi \mathbf{P}$	$\pi \mathbf{Q} = 0$
160	-19	In the next chapter	In the next two chapters
160	-15	(1966)	(1956)
161	Ex. 4.2(b)	$0 < j < n.$	$1 < j \leq n.$
163	Ex. 4.10	(See Errata Sheet Appendix)	
163	Ex. 4.12(a)	**(a)** Draw the ... for the case ...	**(a)** Draw the ... for the case ...
165	-1	$\pi = \pi P$	$\pi = \pi \mathbf{P}$
168	15	$x^k \triangleq$	$\overline{x^k} \triangleq$
179	Fig. 5.3	Chain.	chain.
179	8	$\int_0^\infty P[\tilde{v} = k, x < \tilde{x} \leq x + dx] \, dx$	$\int_0^\infty P[\tilde{v} = k, x < \tilde{x} \leq x + dx]$
187	-2	By definition we have	We have
187	Eq. (5.64)	$\overline{N} \triangleq \sum_{k=0}^{\infty} kP[\tilde{q} = k]$	$\overline{N} = \overline{q} \triangleq \sum_{k=0}^{\infty} kP[\tilde{q} = k]$
190	7	(See Errata Sheet Appendix)	
198	Eq. (5.97)	$E[e^{-s\tilde{s}}]$	$E[e^{-s\tilde{s}}]$
201	16	principle	principal
212	-14	$P[\tilde{v} = k]$	$P[\tilde{v} = k \mid x_1 = x]$
231	15	Intensitatschwankungen	Intensitätsschwankungen
231	16	**6**, 1−189	**44**, 1−189
231	17	dev	der

220

PAGE	LINE	ORIGINAL TEXT	CORRECTED TEXT
231	Ex. 5.2(a)	$P[Y \leqslant y \mid t]$	$P[Y \leqslant y]$
232	Ex. 5.2(c)	(See Errata Sheet Appendix)	
232	Ex. 5.3(a)	$\bar{N}_q, \rho, \bar{x}, \sigma_b^2$ and $P[\tilde{w} > 0]$.	$\bar{N}_q, \rho, \bar{x},$ and C_b^2.
232	Ex. 5.5(a)	(See Errata Sheet Appendix)	
232	Ex. 5.6(a)	$\beta_{n+1} \leqslant k$	$\beta_{n+i} \leqslant k$
233	Ex. 5.7	$P_k(t, x_0) dx_0$	$P_k(t, x_0) \Delta x_0$
233	Ex. 5.7	$x_0 + dx_0$	$x_0 + \Delta x_0$
235	Ex. 5.12(a),(b)	(See Errata Sheet Appendix)	
236	Ex. 5.13(b)	$p[\tilde{q} = k]$	$P[\tilde{q} = k]$
237	Ex. 5.17	We begin at a random time	Begin at an arrival time
238	Ex. 5.17	measure the time w	measure the time $\tilde{w}$
238	Ex. 5.18(f)	from (e).	from (e).
240	Ex. 5.23	z-transform for the number	z-transform for $\tilde{f}$, the number
240	Ex. 5.23(a)	and j	and $\tilde{f}$
240	Ex. 5.23(b)	behind.) (HINT: condition on j.)	behind.)
242	2	implicity	implicitly
244	3	$1 - e^{-\mu_t}$	$1 - e^{-\mu t}$
247	12	through state E_j	through any state E_j
251	-11	(3.24)	(3.21)
254	-5	1,0 in order.	1 in order.
281	16	$\int_{t=0}^{\infty} P[x_n \leqslant u + t \mid t_{n+1} = t] \, dA(t)$	$\int_{t=0}^{\infty} P[x_n \leqslant u + t_{n+1} \mid t_{n+1} = t] \, dA(t)$
282	3	$\int_{0^-}^{\infty} P[u_n \leqslant y - w \mid w_n = w] \, dW_n(w)$	$\int_{0^-}^{\infty} P[u_n \leqslant y - w_n \mid w_n = w] \, dW_n(w)$
282	-8	$C(y-w) W(w) \vert_{w=0^-}^{\infty} - \int_{0^-}^{\infty} \cdots$	$C(y-w) W(w) \vert_{w=0^-}^{\infty} + \int_{0^-}^{\infty} \cdots$
282	-7	$C(y) W(0^-) - \int_{0^-}^{\infty} \cdots$	$C(y) W(0^-) + \int_{0^-}^{\infty} \cdots$
282	Eq. (8.22)	$W(y) = \begin{cases} -\int_{0^-}^{\infty} \cdots \\ 0 \end{cases}$	$W(y) = \begin{cases} \int_{0^-}^{\infty} \cdots \\ 0 \end{cases}$
283	6	Weiner	Wiener
283	16	Weiner	Wiener
299	7	arrival must wait	arrival does not wait
300	9	Keilsen	Keilson
301	12	matrix of the transition probabilities	vector of the equilibrium probabilities

221

PAGE	LINE	ORIGINAL TEXT	CORRECTED TEXT
308	Eq. (8.110)	$\rho \dfrac{(1+C_b{}^2)}{2\mu(1-\rho)}$	$\dfrac{\rho \bar{x}(1+C_b{}^2)}{2(1-\rho)}$
312	9	(See Errata Sheet Appendix)	
313	23	dune'	d'une
313	23	par de	par des
313	-6	SYSK 62	SYSK 60
313	-5	1962	1960
314	Ex. 8.5	$\int_{-\infty}^{y} R(u,y)\, dC(u)$	$\int_{-\infty}^{\infty} R(u,y)\, dC(u)$
315	Ex. 8.6(c)	[SYSK 62]	[SYSK 60]
318	Ex. 8.17(c)	(See Errata Sheet Appendix)	
320	9	Chapter 5 addresses itself	Chapters 5 and 6 address themselves
320	12	The chapter is	These chapters are
320	13	identifies	identify
329	7	$|z| < 1/\alpha$	$|z| < 1/|\alpha|$
329	8	α	$|\alpha|$
330	entry 10	$(n-2),\ldots,(n-m+1)$	$(n-2)\cdots(n-m+1)$
334	Eq. (I.23)	$z = 1/\alpha i$	$z = 1/\alpha_i$
337	12	on and within the closed contour C.	on and within the closed contour C except at these singular points.
339	-2	real number σ_a such that	real number c satisfying
339	-1	$\lim_{\tau\to\infty}\int_0^{\tau} |f(t)|\, e^{\sigma_a t}\, dt < \infty$	$\lim_{\tau\to\infty}\int_0^{\tau} |f(t)|\, e^{-ct}\, dt < \infty$
340	1	The smallest possible value for σ_a is referred to as	The greatest lower bound of the set of such c is denoted by σ_a and is referred to as
341	16	behoves	behooves
345	-9	$\int\cdots\int f(x)\, dx$	$\int\cdots\int f(x)\, (dx)^n$
346	entry 4	$f(t-a)$	$f(t-a) \quad (a \geqslant 0)$
346	entry 13	$\int_{-\infty}^{t} f(t)\, dt$	$\int_{-\infty}^{t} f(u)\, du$
346	entry 14	$\int_{-\infty}^{t}\cdots\int_{-\infty}^{t} f(t)\,(dt)^n$	$\int_{-\infty}^{t}\cdots\int_{-\infty}^{t} f(u)\,(du)^n$
346	†+3	$\int_{-\infty}^{t}\cdots\int_{-\infty}^{t} f(t)\,(dt)^n$	$\int_{-\infty}^{t}\cdots\int_{-\infty}^{t} f(u)\,(du)^n$
353	Fig. I.4	$W = Im(s)$	$\omega = Im(s)$
353	-3	$e^s{}_t$	e^{st}
371	11	Eq. (II.8)	Eq. (II.9)
379	6	$E_Y[y]$	$E_Y[Y]$
383	-12	$\int_0^{\infty} |e^{-sx}|\,|a(x)|\, dx$	$\int_{0^-}^{\infty} |e^{-sx}|\,|a(x)|\, dx$

PAGE	LINE	ORIGINAL TEXT	CORRECTED TEXT
383	−8	$\leq |e^{-\sigma x}| |e^{-j\omega x}|$	$= |e^{-\sigma x}| |e^{-j\omega x}|$
383	−5	$\int_0^\infty a(x)\,dx$	$\int_{0^-}^\infty a(x)\,dx$
385	8	$\sum g_k$	$\sum_k g_k$
398	1	$Rc(s)$	$Re(s)$
400	7	$\rho = \lambda \bar{x}$	$\rho \triangleq \lambda \bar{x}$
405	10	$j = 1, 2, \ldots, r$	$j = 1, 2, \ldots$
407	−1	$Q(z) = \dfrac{(1-\rho)(1-z)B^*[\lambda - \lambda G(z)]}{B^*(\lambda - \lambda G(z)) - z}$	$Q(z) = \dfrac{(1-\rho)(1-G(z))B^*[\lambda - \lambda G(z)]}{\bar{g}\,(B^*[\lambda - \lambda G(z)] - z)}$
407	−1	(bulk arrival)	(bulk arrival at departure instants)
413	7	Interdeparture time distribution, 148	Interdeparture time distribution, 148, 238

ERRATA SHEET APPENDIX

PAGE	COMMENT/CORRECTION
46	Starting at the top of the page, replace the proof of Eq. (2.85) by the following alternative proof:

exponential distribution. Setting $g(t) = P[\tau_i > t]$ we have, by definition, the following relationship:

$$g'(t) = \frac{d}{dt}(P[\tau_i > t]) = \frac{d}{dt}(1 - P[\tau_i \leq t])$$

$$= -f_{\tau_i}(t) \qquad (2.84)$$

where we use the notation $f_{\tau_i}(t)$ to denote the pdf for τ_i. Now Eq. (2.83) states

$$g(s+t) = g(s)g(t).$$

Subtracting $g(t)$ from both sides of this equation and dividing by s yields

$$\frac{g(s+t) - g(t)}{s} = g(t)\frac{g(s) - 1}{s} = g(t)\frac{g(s) - g(0)}{s}$$

where we have used $g(0) = P[\tau_i > 0] = 1$. Taking the limit as $s \to 0$ gives

$$g'(t) = g(t)g'(0)$$

and thus

$$\frac{g'(t)}{g(t)} = g'(0).$$

Integrating with respect to t gives

$$\log_e g(t) = g'(0)t + K.$$

As $g(0) = 1$ we have $K = 0$. So $g(t) = e^{g'(0)t}$ and, since $g'(t) = g(t)g'(0)$, we have

$$g'(t) = e^{g'(0)t}g'(0).$$

We use Eq. (2.84) to obtain the pdf for τ_i as

$$f_{\tau_i}(t) = f_{\tau_i}(0)e^{-f_{\tau_i}(0)t} \qquad (2.85)$$

which holds for $t \geq 0$.

PAGE	COMMENT/CORRECTION

85 — The indices on the right-hand side of the equation in Ex. 2.20(e) are incorrect. The corrected equation is as follows:

$$P_k^*(s) = \frac{1}{\lambda}\left[\alpha_1^{i-k-1} + \left(\frac{\mu}{\lambda}\right)\alpha_1^{i-k-3} + \left(\frac{\mu}{\lambda}\right)^2 \alpha_1^{i-k-5} + \cdots \right.$$
$$\left. + \left(\frac{\mu}{\lambda}\right)^i \alpha_1^{-i-k-1} + \left(\frac{\lambda}{\mu}\right)^{k+1} \sum_{j=k+i+2}^{\infty} \left(\frac{\mu}{\lambda\alpha_1}\right)^j \right]$$

85 — Exercise 2.20(g) is to be added as follows:

(g) Starting with the equation in part (d), extend the applicability of Eq. (2.163) to the range $k < i$.

94 — (lines 2-3) The condition for ergodicity may be replaced simply by $S_1 < \infty$; that is, $S_2 = \infty$ is superfluous.

94 — (lines 8-9) The condition for transience may be replaced simply by $S_2 < \infty$; that is, $S_1 = \infty$ is superfluous.

94 — (lines 11-14) Replace present sentence with the following:

We note that the condition for ergodicity is met whenever the sequence $\{\lambda_k/\mu_{k+1}\}$ remains bounded away from unity from some k onwards, that is, if there exists some k_0 and some $C < 1$ such that for all $k \geqslant k_0$ we have

$$\frac{\lambda_k}{\mu_{k+1}} \leqslant C < 1 \qquad (3.20)$$

163 — Exercise 4.10 is to be corrected as follows:

4.10 We consider the denominator polynomial in Eq. (4.35).
(a) Of the $r+1$ roots, one occurs at $z=1$. Use Rouche's theorem to show that exactly r roots lie in the unit disk $|z| \leqslant 1$.
(b) Show that $z=1$ is the only root on the unit circle $|z| = 1$.
[Hence (a) and (b) establish that exactly $r-1$ roots lie in the range $|z| < 1$, and one root, say z_0, lies in the region $|z_0| > 1$.]

190 — Two sentences (lines 7-10) are logically out of order. They should read:

system at customer departure instants. We know that $\bar{q}$ also represents the average number of customers found at random, and so we may equate $\bar{q} = \bar{N}$. We may therefore apply Little's result to this expected number in order to obtain the average time spent in the system (queue + service).

232 — Exercise 5.2(c) is to be corrected as follows:

(c) Let $\hat{F}(y) = \lim \hat{F}_t(y)$ as $t \to \infty$ with corresponding pdf $\hat{f}(y)$. Show that we now have
$$\hat{f}(y) = \frac{1 - F(y)}{m_1}$$
[HINT: Use the key renewal theorem (see [FELL 66] or [TAKA 62a], for example).]

232 — The second sentence of Exercise 5.5(a) is to be removed.

PAGE	COMMENT/CORRECTION
235	Exercise 5.12(a) and (b) is to be corrected as follows:

5.12 Consider the M/G/1 bulk arrival system in the previous problem. Using the method of imbedded Markov chains:
(a) Find the expected queue size at departure instants.
[HINT: show that $\bar{v} = \rho = \lambda \bar{x} \bar{g}$ and

$$\overline{v^2} - \bar{v} = \left. \frac{d^2 V(z)}{dz^2} \right|_{z=1} = \rho^2(1 + C_b^2) + \rho \bar{g}(1 + C_g^2) - \rho$$

where C_g is the coefficient of variation of the bulk group size and $\bar{g}$ is the mean group size.]
(b) Show that the generating function for queue size at departure instants is

$$Q(z) = \frac{(1-\rho)(1-G(z))B^*[\lambda - \lambda G(z)]}{\bar{g}(B^*[\lambda - \lambda G(z)] - z)}$$

312	(lines 9-13) Replace present sentence with the following:

Now the dual system is an unstable G/M/1; we ask in part (ii) of Exercise 8.17(c) that the reader show for such a queue that the idle period pdf will have transform

$$\hat{I}^*(s) = \frac{1 - B^*(s)}{s\bar{x}} \qquad (8.121)$$

318	Exercise 8.17(c) is to be corrected as follows:

(c) Consider a G/M/1 queue.
(i) For $\rho < 1$ (stable queue), use Eq. (8.106) to show directly that

$$I^*(s) = \frac{A^*(s) - \sigma}{-\frac{s}{\mu} + 1 - \sigma}$$

(ii) For $\rho > 1$ (unstable queue), use (a) to show that

$$I^*(s) = \frac{1 - A^*(s)}{s\bar{t}}$$

which gives Eq. (8.121) as promised.

ERRATA SHEET

for

QUEUEING SYSTEMS
VOLUME I: THEORY
(first printing only)

PAGE	LINE	ORIGINAL TEXT	CORRECTED TEXT
xii	−17	pursue further	further pursue
6	6	size	*size*
9	−7	Ford, L. K.	Ford, L. R.
9	−5	*Transportation, and Transmission*	*Transmission, and Transportation*
17	−11	[JEWE 67] and S. Eilon [EILO 69].	[JEWE 67], S. Eilon [EILO 69] and S. Stidham [STID 74].

225

PAGE	LINE	ORIGINAL TEXT	CORRECTED TEXT
21	-3	X_{n+1}	x_{n+1}
21	-2	X_n	x_n
51	Eq. (2.107)	$q_{ij}(t) \triangleq q_{ij}$	$q_{ij} \triangleq q_{ij}(t)$
52	5	(See Errata Sheet Appendix)	
64	-12	$\lambda \beta_i e^{-\lambda \beta}$	$\lambda \beta_i e^{-\lambda \beta_i}$
73	Eq. (2.151)	$\left.\begin{array}{l}\lambda_k = \lambda \\ \mu_k = \mu\end{array}\right\} \xleftrightarrow{M/M/1} \left\{\begin{array}{l}A(t) = 1 - e^{-\lambda t} \\ B(x) = 1 - e^{-\mu x}\end{array}\right.$	$\left.\begin{array}{l}\lambda_k = \lambda \\ \mu_k = \mu\end{array}\right\} \xleftrightarrow{M/M/1} \left\{\begin{array}{l}A(t) = 1 - e^{-\lambda t} \\ B(x) = 1 - e^{-\mu x}\end{array}\right.$
78	-1	(See Errata Sheet Appendix)	
86	Ex. 2.22	to the Poisson distribution that if	that if
94	$*+1$	both $\tilde{t}$ and $\tilde{x}$ are both	$\tilde{t}$ and $\tilde{x}$ are both
101	12	equivalent, and so we	equivalent for $\alpha = \lambda$, and so we
120	5	as the variance of $b(x)$.	as the service time variance.
125	-4	later.	later or some other distribution.
135	Fig. 4.8	dg_i	λg_i
150	-1	$\binom{N+K-1}{K-1}$	$\binom{N+K-1}{N-1}$
160	-16	BURK 66	BURK 56
209	Fig. 5.11a	I_1	τ_1
230	Eq. (5.181)	$-\lambda \int_{x=0}^{w} B(w-x)\, dF(w)$	$-\lambda \int_{x=0}^{w} B(w-x)\, dF(x)$
282	-5	is the probability than an	is the probability that an
289	-13	customer need not queue†	customer need not queue†.
295	Fig. 8.7	$\text{Im}(s) = w$	$\text{Im}(s) = \omega$
295	9	$\sigma + jw$	$\sigma + j\omega$
375	-9	$P[Y \leqslant y)]$	$P[Y \leqslant y]$
399	10	$a_1, a_2 \cdots a_k$	$a_1 a_2 \cdots a_k$
404	3	$\binom{2n-2}{n-1}$	$\binom{2n-2}{n-1}$

ERRATA SHEET APPENDIX
(first printing only)

PAGE	COMMENT/CORRECTION
52	(lines 5-6) Omit the following sentence:
	In this case our chain is ergodic.
78	Add the following reference at the bottom of the page:
	STID 74 Stidham, S., Jr., "A Last Word on $L = \lambda W$," *Operations Research*, **22**, 417–421 (1974).